The

LONGEVITY
BOOK

———

Also by Cameron Diaz and Sandra Bark

The Body Book

The
LONGEVITY
BOOK

The Science of Aging, the Biology of Strength,

and the Privilege of Time

CAMERON DIAZ

and SANDRA BARK

HARPER WAVE

"Chorus of Cells," reprinted with permission from *Poems from the Pond: 107 Years of Words and Wisdom; The Writings of Peggy Freydberg*, edited by Laurie David.

HarperCollins books may be purchased for educational, business, or sales promotional use. For information, please e-mail the Special Markets Department at SP-sales@harpercollins.com.

FIRST EDITION

Designed by Headcase Design • www.headcasedesign.com

Photo opposite page x © Jeff Dunas

Illustrations by Harriet Russell

Library of Congress Cataloging-in-Publication Data has been applied for.

ISBN: 978-0-06-237518-6
ISBN: 978-0-06-246411-8 (BAM Signed Edition)
ISBN: 978-0-06-246410-1 (B&N Signed Edition)
ISBN: 978-0-06-246427-9 (Indigo Signed Edition)

16 17 18 19 20   OV/QGT   10 9 8 7 6 5 4 3 2 1

Dedicated to your journey

CONTENTS

ALL YOU HAVE TO DO IS TO PAY ATTENTION; LESSONS ALWAYS ARRIVE WHEN YOU ARE READY, AND IF YOU CAN READ THE SIGNS, YOU WILL LEARN EVERYTHING YOU NEED TO KNOW IN ORDER TO TAKE THE NEXT STEP.

—PAOLO COELHO, *THE ZAHIR*

The

LONGEVITY
BOOK

———

PREFACE

THERE'S AN IDEAL BEAUTY THAT IS HARDER TO
DEFINE OR UNDERSTAND, BECAUSE IT OCCURS NOT JUST
IN THE BODY BUT WHERE THE BODY AND THE SPIRIT
MEET AND DEFINE EACH OTHER.

—URSULA K. LE GUIN, *THE WAVE IN THE MIND*

THE BLACK AND WHITE photo you see on the previous page was inspired by a picture taken at the very first professional photo shoot I was ever a part of. I was brand-new to the industry—only sixteen years old—when the amazing photographer Jeff Dunas invited me to sit for him. That first day, he draped me in a white sheet and told me to look at the camera. It was the beginning of a new life for me. Jeff had the idea that we should repeat this image, as many times as we both were able to, for the rest of our lives. I thought it was a brilliant idea and certainly a unique opportunity. So six years later, we recreated the photo, and this time I held the original photo in my hands. And then another six years later, when I was twenty-eight years old, we did it again.

While I was working on this book, considering the meaning of my life's journey, it reminded me that Jeff and I had made this pact and that it had been way too long since we had taken a photo. So I called him up and soon I was once again in front of his lens. The photo you see in this book is the result.

With every passing year, my physical body has shifted, of course—that's easy enough to see in the picture. But less obvious are the shifts that can't be weighed or measured; the emotional, mental, and spiritual shifts that accompany the passage of time and the accumulation of experiences. If you squint really hard, if you look really closely, you might be able to see some evidence of those changes. When I look at this picture, I can see them and feel them immediately.

Contemplating aging has a way of making us consider our youth. In order to look forward, we first look backward. We flip through old photographs and read through old letters and journals. We reminisce with friends and family about the experiences that have led us to where we are today. Sometimes we feel nostalgic. Other times we feel relieved that no matter what, time keeps marching on.

When I look at this collection of photographs, I feel the pull of time, forward and backward, each image bringing me closer to the woman I am today, each pointing toward the woman I will become. I don't yet know who she is, but I look forward to meeting her. We all want to know what, when, and how our stories will unfold, and all we have to go on is the life we have experienced so far, the choices we have already made, and the stories we have lived through, from beginning to middle to end.

I wrote this book because I wanted to peek into my future. I wanted to get a sense of what might happen, what could happen, and what I could possibly do for myself now to continue the journey, and to enjoy the journey for as long as possible. In the years that come, I may grow weaker, but it is my hope that I can also grow wiser, warmer, and more resilient. I hope we can all find the power to grow older together, each of us doing the work we must to become stronger and more loving and more at home in our hearts, in our bodies, and in the world.

Photographs make it possible for us to watch ourselves age. We can see ourselves grow taller, observe the cheekbones that show up as we pass from adolescence to young womanhood, notice the wrinkles that begin appear just a couple of decades after that. What is less easily grasped by a camera lens is the inner growth, the way the heady passions of youth grow into the steadier fascinations of adulthood, or the privileges that time offers with every passing year.

Time can be kept by clocks and calendars, measured in inches and wrinkles, and caught in images and photographs. But if we are very lucky, it can also be counted in a life well spent, full of learning, love, and laughter.

INTRODUCTION

ROUND MY FORTIETH BIRTHDAY, I started thinking about what it means to age. It is a fundamentally human question, one we all start to consider at some point. None of us is immune to the passage of time, and one day, when you realize that life just keeps moving forward and there really is no going back—the wondering begins. Poets write poems about it and musicians write songs about it and scientists design experiments to understand it. All of us humans wonder what will happen to us when we get older.

I had been living in this body of mine for more than four decades when I started thinking about the changes that might be coming down the road. I have experienced a lot of changes throughout my lifetime, of course, but I found myself unable to stop thinking about how the decades ahead were going to reveal some even more drastic changes—and how I didn't really understand the aging process, or what it would mean for me. I had seen people I love get old and decline sharply and painfully, and I wondered if that would be my fate, or if I could hope for something better.

Around the same time, I was also writing a book called *The Body Book*, which focused on the foundational aspects of human life. It was full of the kind of stuff I had been learning about over the course of nearly two decades—information about nutrition, exercise, and cultivating strong habits—along with some of the latest scientific insights about overall physical health. I already had intimate knowledge of the ways in which fitness and diet could change my body for the better. Now I wondered: how could I stay healthy and strong in the years to come? We all want longevity, of course. We all want more calendar pages to turn, more time to experience life. But what is a long life without strength, without physical and emotional health and resilience?

So I called my writing partner, Sandra Bark, with whom I had written *The Body Book*, and I told her that I had figured out that our next book would be about cellular aging.

She laughed and said, "Great, an easy one."

To be clear, there is nothing easy about this subject of aging—not the science of it, and not the experience of living through it. But easy or not, it will happen, and it is happening right now. We can avoid most uncomfortable truths for a very long time, if we want to, but there's no denying that this one catches up with us eventually. It's my hope that with a better understanding of what aging really is—the science of it, the biology of it, the cultural and historical context of it—we can all become empowered to live well in the years ahead.

This is not an antiaging book. I don't want you to live in fear of aging . . . I want to reframe the way that we, as women, talk about aging.

One thing that I've learned about uncomfortable truths is that you make life a whole lot harder for yourself when you pretend they aren't real. You can waste a lot of precious time and energy trying to make something into what it is not. Once you stop fighting reality, everything becomes a lot easier. Youth is a beautiful part of life, and the discoveries we make when we are young are invaluable. They are the lessons and the memories that we will carry with us as we move into each new phase of our lives. It's important to keep those lessons close to us, but it's also important to let go of what no longer is, and to accept and prepare for what is to come.

As babies and toddlers, we were blissfully unaware of the fact that we were zooming ahead developmentally. As adolescents transitioning into teenagers, we were equipped only with the information the adults around us decided to share (for better or for worse), and our understanding of what was around the bend and how to deal with all the crazy changes we'd soon experience wasn't up to us. This round, it's our turn. When it comes to the next phase of our lives, the responsibility of preparation is solely ours. We have the opportunity to gather our resources, our abilities, all the wisdom we have gained over the years, and design a plan for healthy aging that will help us stay strong while also making us more aware, more conscious, and more connected to ourselves and to one another.

Before we embark on this journey together I would like to offer a disclaimer: This is not an antiaging book. I'm not going to tell you how to trick time or reverse the aging process in thirty days. Some books and articles about aging claim that the latest groundbreaking discoveries can show you how to turn back the clock. Others offer strategies for making yourself *look* younger, or suggest that certain miracle foods or supplements are the newest fountain of youth. This is not that kind of book. This book takes a step back, to examine how the aging process really works and how time will affect us physically and emotionally—because these two components of our health are inseparable.

What you will find in these pages is information and an ideology that I hope will help you find a new way of thinking about aging. I don't want you to live in fear of aging, or beat yourself up about the fact that your body is doing something totally natural. I want to reframe the way that we, as women, talk about aging. I want to offer a perspective that is healthier and more scientifically accurate than the fear- and shame-based conversation that permeates our culture.

What I want for you, for me, for all the women I care about—those I already know and those I haven't yet met, those who are crossing the threshold into middle age now and those who are following behind us—is to be able to approach this subject with knowledge and with confidence instead of sheer terror and a heavier hand with the foundation. And by "knowledge," I mean having the facts to live better, longer, and stronger. And by "confidence," I mean having the ability to own our age instead of hiding from it or apologizing for it. I'm not saying that aging isn't scary. It is. But we can prepare ourselves now for the changes that lie ahead.

I also want you to feel empowered to participate in the new conversation about aging that is turning up everywhere. From public and private funds for scientific research to articles to podcasts to books like this one—everyone is curious to learn more about how we can age better. Part of the reason there's such a sudden flurry of interest in how to age well is because this topic of aging is still so *new*. As you will learn in the pages ahead, at this particular moment in the history of human evolution, our life expectancy is longer than it has ever been. Our relationship with and our understanding of the aging process is still unfolding.

The newness of aging makes the exploration of this topic all the more challenging and all the more thrilling. So when Sandra and I set off on our journey, we went in with open, inquisitive, and studious minds. We talked with researchers and physicians and educators, and visited universities and research centers like the National Institutes of Health (NIH). And to our surprise we learned that although human aging is a relatively new phenomenon, the most scientifically vetted, cutting-edge ways of staying healthy and strong are actually not very new or complicated at all.

In fact, the best things we can do for ourselves as we grow older also happen to be some of our favorite things to do. Eating good food, developing our muscles, getting a good night's sleep, loving other people, laughing, relaxing, finding joy in the world. These are the actions and activities that make us interesting people, curious people, strong people. Who doesn't love a good meal with dear friends, or embracing a loved one? Who doesn't love to laugh her ass off, or go for a long walk, or have a new adventure? How about taking a few moments to breathe deeply and let the cares of the day slip away? How nice does that sound?

To us, it sounds like a revelation. The best way to age well isn't to worry about aging. It is to live well.

Today we have a deeper understanding of how our body functions on a cellular level than ever before, and because of that, we can see how things like food, movement, rest, meditation, social connection, learning, and the overall enjoyment of life serve to make us stronger and healthier deep within our cells. That's right—having a laugh has an actual impact on your cells. Spending time with good friends is beneficial for your cells. All those elements that make life beautiful and wonderful are *good* for you.

That's why we wrote this book. To share the science of aging. To provide the information you need to make choices that support your health as you age, which can help slow the rate at which you age, and in some cases, repair damage that has already been done. To help you understand the conversation about aging, which is getting more and more robust each day. We are all aging, you and I, and the sooner we come to terms with that, the more opportunities we can give ourselves to age with health and with joy.

The good thing about this journey is that even though the road ahead is unknown, you're still traveling in the same direction: deeper and deeper into the depths of *you*. Along with signs of aging, don't you see signs of growth? Are you a stronger person than you were a decade ago, more knowledgeable, more in tune with who you are and what you need and who and what you love? Life comes with some sharp curves, and every journey has a few missed turns along the way. But there are also the scenic overlooks, where the horizon suddenly opens up, and you can admire the view and appreciate the hard work it took to get there.

Having a laugh has an actual impact on your cells. Spending time with good friends is beneficial for your cells. All those elements that make life beautiful and wonderful are *good* for you.

Appreciating all the ways we can evolve over the years—the self-knowledge we develop, the skills and wisdom we accumulate, the relationships we build with others and with ourselves—these are the privileges of time. There's no denying the decline that accompanies aging. But growing older also offers opportunities. The idea we can grow stronger as we age—it feels good to me. It feels right. It feels possible.

And the new science of aging backs that up.

PART I

THE SCENIC ROUTE

Living in the Age of Longevity

THE PRIVILEGE OF A LIFETIME IS BEING WHO YOU ARE.

—JOSEPH CAMPBELL

CHAPTER 1

BEAUTY IS TIMELESS, WISDOM IS PRICELESS

What We Talk about When We Talk about Aging

N A BUSINESS THAT is obsessed with youth, I am no longer considered a young woman. This was made clear as soon as I hit the ripe old age of thirty-nine. I can't tell you how many times a journalist asked me if *as an actress,* I was scared to turn forty. As these questions about my age seemed to become a consistent part of every press interview, I realized just how frightened we all are of getting older. We make jokes about it, or we see it as sad, as ugly, as dangerous.

The conversation we have around aging in our culture feels very misplaced to me. Am I afraid to turn forty? These people who were asking about my age in front of a camera weren't wondering if I was afraid that my health might decline after forty. They weren't concerned that my organs might experience the effects of aging. They weren't asking what aging means to me, as a woman, as a human being, as a living organism with an expiration date.

They were saying, "Aren't you afraid that the death of your career is imminent because you don't *look* twenty-five anymore?"

The funny thing is, those people who suggested that I'd reached my expiration date at an age when I still felt pretty damn good were actually doing me a real favor: they were jump-starting my thought process about what aging is and what kind of impact it will have on me. The conclusion I came to was that as long as I get to keep on aging, I'm pretty lucky. Not everyone has the opportunity to grow old. Some people die before they have a chance to celebrate another birthday.

So to answer the questions those journalists asked about how my opinion of myself has changed as my looks have changed, my answer is that aging is a privilege and a gift. As we get older, I believe beauty appreciates, not depreciates. It grows, not fades. With age, I have developed a more nuanced understanding of what beauty really is. Beauty is not just something you are born with. Beauty is something you grow into.

As I start this next phase of my journey, I feel proud of where I've come from and curious about what's ahead. I don't know what life will hold for me. But I am ready. Because I know myself better than I did years ago, and I trust myself to make good decisions, or at least to do my best. Because I value the lessons that I've learned, especially in the last decade, and I look forward to seeing what kinds of new understandings the decades ahead will bring.

Beauty appreciates, not depreciates. It grows, not fades. With age, I have developed a more nuanced understanding of what beauty really is. Beauty is not just something you are born with. Beauty is something you grow into.

WHERE DID YOU LEARN ABOUT
BEING BEAUTIFUL?

My first model of beauty was my mother. I don't think I'm being partial when I say this: my mom is a beautiful woman. She has always had full lips, glowing skin, and blue eyes with a depth of gray that draws you in. She possesses the kind of beauty that shines from the inside out. So as far as I was concerned, she never needed any makeup, but like most other women, she had a "face" that she would apply daily. She would highlight her eyes, brighten her cheeks, and lengthen her lashes. She was so skilled in her routine that it took her exactly the same amount of time every morning to complete it, and her face always looked exactly the same after she had finished. What was even more impressive to me was how subtle but effective her application was at complementing her already luminous beauty.

When we were little, my sister and I loved watching our mom go through this routine and couldn't wait to be old enough to learn how to do it ourselves. And once we were finally old enough—man, we really went for it. Subtlety may have been my mother's gift, but there was little of that in our technique. There were many times when it would have been challenging to distinguish my sister and me from a pair of peacocks. It took years before we learned to refine our hand and apply our "face" a bit less liberally, and even more years before I understood what the point of this ritual really was.

Now I know that adornment is a natural instinct. All over the world, men and women alike invest in beauty rituals to make themselves more attractive. In the Serengeti, Masai warriors spend days decking themselves out in tribal gear, adorning themselves from head to toe with vibrantly colored jewelry and clothing. They paint their faces and braid their hair in elaborate weaves. Some of this decoration serves as an indicator of each man's position in the tribe, and some is simply for beauty's sake—but in either case, the goal is to stand out from the crowd and attract a woman. It can take a warrior and a companion a week to apply the embellishments. A week! That's a pretty significant amount of time for a man who's also in charge of keeping his family's livestock—and his family—safe from predators.

Why am I talking about the beauty rituals of men in a book meant for ladies? Because they help us understand that the desire to look beautiful, the drive to stand out, isn't restricted by age or culture or gender. In fact, it's not even restricted to humans. Animals also possess an instinct for visual attraction, as with the infamous peacock, the spirit animal of my earliest makeup attempts. Richly hued flowers flirt with insects who might spread their pollen near and far. Wanton trees and vines entice animals with beautiful, ripe fruit so the seeds can be dispersed. All of us, from birds to bees to humans, are hopelessly attracted to bright, shiny colors, which is why nature uses them to such great effect.

In the animal and plant kingdoms, beauty is an evolutionary imperative, but when it comes to humans, it is about so much more. Clothing and adornment and makeup can be part of a personal narrative, can be about belonging, about blending in, or about standing out. Beauty is an instinct we all share, but our definitions of what is beautiful and our expectations for ourselves and others are shaped, in part, by cultural and social values.

Throughout our lives, we are exposed to ideas about what beauty is, how important it is, and what we can do to make ourselves more beautiful. When we are young we are receptive to those ideas almost without realizing it.

THAT YOUTHFUL GLOW

When my sister and I were fifteen-year-olds painting on hot-pink lips and sparkly blue eyes in an attempt to look older, we were oblivious to the fact that most women actually apply makeup to look younger. When it comes to fifteen-year-olds, nature is in the habit of generously handing out rounded cheeks still plump with baby fat, bodies unaffected by gravity, and shiny, silky hair. Of course, as self-conscious teenagers, we never would have thought of ourselves as beautiful.

In fact, few women seem to fully recognize the attributes they possess when they possess them. I think we've all had the experience of looking at a picture that was taken ten years ago and thinking, "Wow! I was so young and pretty back then. But I know I didn't feel that way when the picture was taken. Why didn't I realize how great I looked?" The truth is that you didn't appreciate how great you looked then because you weren't thinking about where you were on your journey through life. That moment when you thought you looked "old," when you were twenty-five or maybe thirty-five, is the same moment you are experiencing right now, when you are both the oldest you have ever been and the youngest you will ever be. And in ten years, when you look at a photo of yourself that was taken today, you will notice how young you look, and wonder why you didn't realize it then. It's just what we do.

When you possessed those attributes of youth, you also probably didn't think about the fact that that they wouldn't last forever. You probably couldn't imagine that one day you would notice that your skin wasn't as smooth as it used to be, or that the hair on your head was becoming less lustrous, or that random hairs were cropping up in strange places. You couldn't possibly envision that your body would find itself on the losing end of gravity at some point. But there comes a time in all our lives when we become aware that we are starting to age. We find our first gray hair or notice laugh lines in the mirror that seem to have appeared overnight. And at that moment you might ask yourself: *What the hell is going on?!*

In the plant and animal kingdom, beauty is an evolutionary imperative, but when it comes to humans, it is about so much more.

Well, dear female friend, what is happening to you is also happening to every living organism on the planet. Because all living creatures age. The process can take one day, as in the case of the mayfly, or as long as 250 years, as with giant land tortoises. As soon as we are sexually mature enough to reproduce, aging happens to us all. With each day that goes by, imperceptible changes are taking place within our cells, and as the decades accumulate, those changes begin to show up as streaks of gray, and as wrinkles, and in a lot of other superficial ways.

The desire to be forever young is not a modern-day preoccupation—just ask Ponce de León. Antiaging procedures have been around for millennia. Some are gross. Some are weird. Some rely, ironically, on dead people. And some, even more ironically, will actually kill you. Here's a brief history of the antiaging industry.

Circa 70 BC:	Ancient Egyptians:
Cleopatra reputedly enjoyed facial masks made of readily available crocodile poop from the Nile.	Used eye pencils made of lead, a heavy metal linked to skin diseases, infertility, and death.

1513:	Circa 1600:
Ponce de León set out to find the fountain of youth. Ended up in Florida (currently the state with the highest percentage of elderly people in the United States) instead.	In the kingdom of Hungary, Countess Elizabeth Bathory reportedly bathe in the blood of virgins to maintain her youthful glow, giving rise to centuries of vampire legends.

1906:	1992:
Congress passed the Pure Food and Drug Act, which said that Americans must stop putting poisons (like lead) on their faces.	Botox was introduced as a treatment for brow wrinkles, and we began to inject poison *into* our faces instead of simply applying it *on* our faces.

Ancient Greeks:

Sought youthful skin through application of white face cream laced with, you guessed it, lead.

Ancient Romans:

Relied on the ammonia in urine to whiten their teeth.

15th-19th century:

Europeans learned nothing from the fall of Rome and sought fairer complexions by using poisonous creams made of lead—because dying young is a great way to stay young forever.

1905:

Surgeons began to offer skin-tightening procedures whereby they made a couple of facial incisions at the side of the face and tugged the skin back. Voilà: the facelift was born. The first textbook on the subject was published a year later in Chicago.

2010s:

A predilection for exotically sourced face masks emerged, a throwback to Cleopatra's reptile-excrement treatments. Bee venom and placenta face masks can easily be purchased online.

2015:

A woman in the UK said that she will quit smiling for forty years in an attempt to avoid getting wrinkles. Other women laughed about this.

There are plenty of methods to make ourselves look like we've shaved off a few years, of course. American women spend $30 billion a year on cosmetics, and I am no different from most women. I've applied makeup to my face for more than a quarter century. I've spent hours in salons getting my hair colored and cut. I've visited my fair share of dermatologists' offices exploring their antiaging arsenals, from creams to lasers to Botox and fillers, all for the sake of maintaining a look of youth and beauty. Many beauty products and procedures really do live up to their promise. They make us feel a little shinier, a little plumper, a little smoother, a little bit better about ourselves. They help us look younger on the outside, which can make us feel like we are younger on the inside. There's nothing wrong with that.

But while a fancy treatment can make you look like you've gotten eight hours of sleep a night for the last decade, your cells know the truth about how you've been spending your waking and sleeping hours. Looking younger is not the same thing as "antiaging." The ability to color our hair and smooth our skin doesn't change the fact that every part of our body is aging every single day.

Believe me, I know that it's easy to get caught up with what you see in the mirror and use it as a metric for how well you are aging. But don't be fooled: just because you look younger than your friend doesn't mean your body isn't experiencing some wear and tear. This aging thing is a process, and we all have our own individual journey through it. What I'd like for you to be aware of as you take that journey is that aging isn't just about your face (or your neck, or your upper arms, or your hands, or . . .). It's about your whole body. And how you take care of your whole body will affect each and every one of your parts, inside and out.

THE NEW CONVERSATION ABOUT AGING

I made my career in a business that must bear a large part of the responsibility for how we, as a society, view aging—a business that tells us that older is ugly or older is less valuable. The message is screaming from every elevation. Think about all the places we see it each day—in magazines, at bus stops, in store windows, on billboards, and in our homes via television or the Internet.

Everywhere you look, the signal is broadcast to women loud and clear: act fast, buy now, change who you are so you don't succumb to the ravages of age. Do not, under any circumstances, let yourself get older.

We've been getting messages from society about how to look for our entire lives. Even teenagers and young women are sent plenty of messages about how they should and could be more attractive. Women of all ages are bombarded with ideas about a standard of beauty that make them feel lousy or as though they have to be different. But with age it gets even more challenging, because these messages begin to suggest that we should actually *be* younger than we are, which is impossible. How can anyone feel good about that?

The physical reality of aging is going to present a real, true challenge to all of us one day. The external signs of getting older are one part of the conversation, but they are not the whole conversation. Societal pressures that encourage women to deny aging or pretend that it's not happening—as though we should somehow be immune to the passage of time—make it an even more painful challenge.

I think it is possible to change the conversation around women and aging—and it starts with conversations like the one we're having here. Instead of whispering about each other for not looking twenty-five, let's encourage real and open dialogue about what we're feeling, what we're wondering about, what we're afraid of, what we're hopeful for. Let's agree to put more value on being a better mother, daughter, sister, wife, friend, colleague, mentor to those around us, instead of acting as if those accomplishments are less important than having smooth skin and a perky bum.

There are many ways to make yourself look younger, but from what I've witnessed among the women in my life, the only way to actually *feel* younger is to embrace the reality that you are in fact getting older—and deal with it. Teenagers look different from toddlers, women in their fifties look different from women in their twenties. That is healthy. That is normal.

So I'd like to propose another message: I'd like to suggest that we all agree, as a group, that every age a woman passes through has its own beauty. Let's raise our standards of beauty and remember that learning and growth and kindness are what we truly value and appreciate in our friends and our sisters and our mothers—and ourselves.

We do not need to look like the images that we are bombarded with at every turn. We do not need to accept the faulty messaging behind those polished-up pictures. We can choose our own role models, women who inspire us to be our best, not someone else's best. We can be the healthiest, most vibrant version of *ourselves* that we can be.

AN APPRECIATION OF TRUE BEAUTY

For years now, I have been painting on different versions of my face in my own beauty routine. Each variation has reflected a different standard of beauty, of what I thought made me attractive to the world at that time in my life. With age, I realize, I have had an opportunity to refine not only my skill with an eyeliner pencil but also my ideas about what makes us beautiful.

Now when I say a women is beautiful, do I mean that she has good bone structure, bright eyes, coiffed hair, muscles that show she works out, curves like a racetrack? Maybe. More likely I mean that she is vibrant, that she is energetic, and exudes an understanding and an acceptance of herself and the world around her. As I make my way toward fifty, I want to earn the next milestone of my life. I want to have lived and learned something new every one of those days that got me there. I accept that I won't be the same person at fifty that I am today. But I hope that I will be wiser, stronger, more compassionate, more conscious of the world around me. Those are the images I want to focus on when I visualize myself growing older—not this compulsion toward youth, toward yesterday, toward a picture of myself that I will never be again. That is my vision for my life. Not looking backward at what I used to have. Looking forward at what I might grow into.

I was at a gathering of family and friends recently, and the women in attendance spanned ages and generations. There were infants and toddlers and children, eleven-year-old girls with knobby knees and their slightly less awkward teenage sisters, and women in their twenties, thirties, forties, fifties, sixties. I couldn't help but think how each woman's beauty was different, distinct. Their smiles were all unique, the color of their skin, their hair, the way they gestured, the way they draped their arms around one another's shoulders or laughingly passed forks and napkins around. There was so much beauty, and none of it had anything to do with age. It had to do with the light that shone from each individual person, from their way of seeing the world and their way of being in the world.

One of the women there was someone I've known since I was sixteen and she was seventeen. She had been a gorgeous girl, and she is still gorgeous today. And looking at her in the sun, I suddenly wondered about how amazing it would feel thirty or forty years from now to know her still, how excited I would be to have made this incredible journey through life with her, women grown up from the girls we used to be. My mother was there, and she also looked so beautiful, and it wasn't because of her makeup skills (even though her blue eyes still look great with a bit of framing), but because of the way she smiles and makes everyone around her feel calm and cared for. She is a beautiful woman because her nature is kind and generous, loving, grounded, and authentic in who she is.

More and more I'm finding that the circle of women around me aren't relying on procedures to help them "appear" to be younger. They are women who are engaged in maintaining the well-being of their mind, body, and spirit. Some of them are fit and full of energy, some have the shine and sparkle of youth, and some have a wicked sense of humor that keeps them laughing at life. But what they all possess in spades is an acceptance of the journey, with all of its unpredictability. Their vitality comes from that embrace, and they meet each new challenge with all the accumulated wisdom they have earned over the years. They have become the women that they were always meant to be.

That's true grace. That's true beauty.

CHAPTER 2

HOW YOUR LIFE GOT LONGER

The Story of Longevity

FEW YEARS AGO MY friends Judd Apatow and Leslie Mann made a very funny movie called *This Is 40,* about a husband and wife dealing with midlife crises and the marital issues that ensue. Part of what makes the movie so great is that it captures, both with humor and poignancy, a popular theme in our culture. Everyone is familiar with the idea of the midlife crisis—that post-forty struggle between accepting that you're getting older and still wanting to remain young and relevant. But the interesting thing about the term is its prefix—"mid-," as in "middle." Because "mid-" implies that we are all bold enough to assume that our forties are the middle of our lives.

While we may feel that turning forty means we're getting old, the truth is that forty used to actually *be* old. Really old. It wasn't "the new thirty"; it was pretty much geriatric. Because in 1850, the life expectancy for a woman in the United States was about forty years old. Less than two hundred years later, that figure has doubled. *Doubled!*

I have to say—I was pretty shocked to learn that only a couple of generations ago, women my age were considered elderly. Forty used to be the end of the line. Nowadays it's more like a springboard for professional advancement and family building and new learning and personal development. Today, people in their forties are working their butts off and building careers and nurturing relationships and starting families while training for marathons and learning how to grow their own herbs and make their own jam. Or they have raised a family and are starting the second phase of their lives, free of the responsibility of child rearing, allowing them to focus once again on their own development.

We are among the first generations to lay claim to our forties as an extension of our thirties instead of a preamble to our seventies. Many of my friends in their forties make jokes about getting old, but they will be the first to admit that they feel, overall, pretty damn young. This does not mean that we are immune from the subtle reminders of aging. Our bodies start to communicate with us in new ways, and sometimes with different needs. I am well aware that the wiggle room I had when I was thirty-five will not be the same when I am forty-five. I try to be constantly in check with the daily choices that make up the equation of my well-being, noticing where there's room for improvement or where I might have room to indulge a little bit more. These days I see

While we may feel that turning forty means we're getting old, the truth is that forty used to actually *be* old. . . . We are among the first generations to lay claim to our forties as an extension of our thirties instead of a preamble to our seventies.

and feel the impact of those choices, for better or worse, on my skin, my flesh and muscle, my energy, and my mood a lot more quickly than I used to. But I've still got a lot of energy, and strong muscles, and I work hard to keep those going. And I know that discipline and the strength it builds are the assets I will need if I want to age with health.

I will continue to work hard because I know that this opportunity to live through the decades, let alone to keep learning and growing with the decades, is new. If forty was still the end of life instead of the beginning of a new phase, I would have never gotten the chance to experience marriage. I would be six feet under instead of planning the next forty years with the love of my life. I think it's so sad that instead of applauding our birthdays, instead of appreciating them, instead of being grateful for this extra time, so many of us lie about our age. As women, we are routinely shamed for aging. We are made to feel like getting older—and especially, looking older—is somehow a personal failure.

When Sandra and I learned that 165 years ago, our sisters had one foot in the grave at forty—our age!—it changed the way we thought about our midlife. That fact that we can grow old enough to look old, in droves, is far from a failure. It happens to be the end product of arguably the biggest success story in human history.

THE MIDLIFE ~~CRISIS~~ CELEBRATION

In order to really appreciate why our midlife crises should actually be midlife celebrations, we need to take a step back and look at the whole of human history as well as our own personal and familial stories. Medicine has made some huge advances over the past century, and all our lives reflect those benefits. Many of us walking around today might not be alive had we been born a hundred years earlier—including me, and including Sandra.

> That fact that we can grow old enough to look old, in droves, is far from a failure. It happens to be the end product of arguably the biggest success story in human history.

When I was only three months old, I woke up with a slight fever. My mother called the doctor, who said that she should continue to monitor me throughout the day. My temperature kept rising. By the middle of the day, my fever was so high she didn't need the thermometer to know something was very wrong. She called my father to tell him to meet us at the hospital. Once we arrived, I was diagnosed and treated quickly. The ER physicians gave me some medicine, my parents took me home, and I recovered within a few days. When Sandra was a child, her mother noticed that she had a strange rash: her body was covered with a smattering of tiny red dots. They rushed to the pediatrician, who diagnosed scarlet fever and gave her some antibiotics. A few days later, she was fine. It was our great luck to be born in an age when drugs like penicillin are easily obtained. But before modern medicine, children died regularly from fevers.

And how about those less dramatic events, the daily occurrences we barely even notice? If I get a cut, I don't think too much about it. I wash the wound with hydrogen peroxide, give it a consistent slathering of antibiotic cream, and it's as good as new by the next weekend. But without those over-the-counter antibacterial and antibiotic helpers, life-threatening infection could set in. People used to die from scrapes, but you and I are confident that we'll be just fine without having to give it another thought.

We all have stories. Scraped knees. The bugs you caught from the neighborhood kids. All those earaches and sore throats. Minor maladies, eased with a trip to the doctor. Our lives are routinely saved by pills and ointments and injections. But for most of human history, infections from cuts could lead to blood poisoning. Illnesses like pneumonia and strep, even diarrhea, could be life threatening.

Today, in the Western world, diseases that were a dire threat since people have been keeping written records have been virtually eradicated. Illnesses that killed kings and queens don't trouble us at all. The tiny blip of history in which we are currently living is the only one in which fear of contracting smallpox doesn't govern our daily activities. And the reason we are granted this good fortune is because regular people and doctors alike got curious about how we could live better, investigated their environments, and then applied what they learned so that we could all become healthier.

That is why aging is, in many ways, a modern phenomenon. The fascinating thing is that we have almost exactly the same DNA as the people who preceded us, but we get to live a lot longer. Genetically speaking, even in ancient Rome, a person who managed to get the right nutrition and get enough sleep and steer clear of diseases and wars and lions and gladiators could have made it all the way to a ripe old age. But the environment of ancient Rome made celebrating your eightieth birthday a pretty impossible goal to achieve. And, in fact, so did most environments of most places, until the twentieth century.

For millennia, smallpox was public enemy number one. It was recorded in the ancient medical literature of Persia, India, and China and has been noted on the remains of Egyptian mummies. Smallpox is even thought to have contributed to the first decline of the Roman Empire. When the Crusaders marched over the continent, smallpox marched home with them. And later, when Europe caught discovery fever and sent explorers and emissaries sailing over the seas to the new lands to the West, smallpox hitched a ride.

In England and in the new American colonies, the death toll was enormous. A bad case of smallpox meant a 60 percent chance of death. By the eighteenth century in England, 400,000 people a year were dying of smallpox, and there was no cure in sight.

But in countries like China, Turkey, and Africa—which had been suffering the ravages of smallpox for eons—a traditional "folk" treatment was helping to stem the tide. In China, a process called inoculation had developed as early as 1100, when it was observed that those who survived smallpox became resistant to the disease. Healers began inserting smallpox-infected needles into otherwise healthy people to deliberately make them sick. This treatment not only helped people survive their case of smallpox, but granted them immunity from future exposure.

Turkish tradesmen had attempted to tell the Europeans about this process, but in England, they weren't buying it. Variolation needed a champion in order to be accepted by Western medicine, and it found one in a woman named Lady Mary Montagu, the wife of England's ambassador to Turkey.

In 1715, Lady Montagu traveled with her husband to Constantinople, and her visit changed the world. In Turkey, Lady Montagu learned about the local method of managing smallpox.

"There is a set of old women who make it their business to perform the operation every autumn," she wrote. "The old woman comes with a nut-shell full of the matter of the best sort of smallpox, and asks what veins you please to have opened. . . . She immediately rips open that you offer her with a large needle . . . and puts into the vein as much venom as can lie upon the head of her needle."

Lady Montagu had an intimate relationship with smallpox. It had killed her brother, and it had left her face badly scarred, a daily reminder of what could happen to her children if they contracted the disease. She wanted to do whatever she could to protect them.

First, she had her young son inoculated by the embassy doctor. Then she brought the procedure back to England and had the same physician inoculate her young daughter in front of an audience of court doctors. Eventually the procedure became widely accepted—a precursor to our modern-day vaccinations.

FIGHTING AN INVISIBLE ENEMY

The first breakthrough in extending the human life span was learning to recognize what was invisible: the microbes that swarm around us—in the air, on our skin, in our food—and have the potential to make us sick. Bacteria and viruses are both types of microbes, invisible to our eyes without a microscope and easily transmitted from organism to organism or via food, water, or air. Basically, everywhere you go, everything you eat, everyone you touch, there they are. Microbes.

Bacteria are tiny one-celled living organisms that have a remarkable ability to flourish in inhospitable conditions, from the iciest regions on earth to the hottest vents beneath the ocean floor. Some bacteria are pathogenic, or capable of causing disease, and have the potential to make us very sick; bacteria cause illnesses like cholera, tuberculosis, and gonorrhea. Many harmful bacteria can be killed simply by washing your hands with soap and warm water. But before the middle of the 1800s, people only washed their hands when they looked dirty—not because they suspected that an invisible bacterial army could really ruin their day.

Until the modern era, in the war between humans and microbes, the microbes were winning. In order to push the limits of life expectancy, doctors first had to understand that much of human illness could be traced to the bacteria and viruses that creep onto and into our bodies. To create medicines that killed germs and saved lives, they had to see what had been unseen.

And once they figured out all of that, there was another battle to win: they had to convince people that they were right.

THE REVOLUTIONARY IDEA OF WASHING YOUR HANDS

In mid-nineteenth-century Europe, one in two hundred women who had a baby did not survive the year after giving birth. It was basically taken for granted that after childbirth, a lot of women would catch something called puerperal fever, an infection of the reproductive organs that often leads to death.

One physician wanted to know why. Ignaz Philipp Semmelweis, a doctor at a Vienna hospital in 1860, noted that in the hospital's two delivery rooms, the rates of infection were skewed. Women in the delivery room where medical students assisted births were three times more likely to come down with puerperal fever than the women in the delivery room staffed by midwives.

Though the chief of his hospital urged him to leave the matter alone, Dr. Semmelweis wanted to investigate the discrepency. Why were so many women and babies dying? Why did more women survive when their babies were delivered by midwife?

Dr. Semmelweis figured out that while the midwives did have better training and more experience than the students, skill wasn't the only reason the

LONGEVITY VOCABULARY

- **LIFE EXPECTANCY**: How long you can be expected to live if you are born in a certain time and in a certain environment
- **LIFE SPAN**: How long an individual actually lives
- **MAXIMUM LIFE SPAN**: The longest recorded life span for the species (122.5 years for a human female)
- **HEALTH SPAN**: The healthy years of your life
- **LONGEVITY**: How long you can live
- **STRONGEVITY**: How strong you are over the course of your long life

patients survived. Remember, nobody understood how germs spread yet. The medical students weren't thinking twice about where they were before they stepped into the room to help deliver babies. But Dr. Semmelweis noted where they were: in class, dissecting diseased cadavers. That's right. The students would go straight from the dissection room to the delivery room *without washing their hands*. They were transmitting bacteria, and ultimately, infection.

When Dr. Semmelweis instructed students to wash their hands before each examination, the maternal and infant mortality rates from puerperal fever dropped sharply. Once one of the top causes of postpartum infection, this disease is now barely seen in the developed world.

Dr. Semmelweis is credited with introducing the idea of antisepsis: sanitizing your hands and keeping surfaces free of germs. His discovery changed our entire understanding of germs and illness. This was a crucial shift for the human life span, because it gave more women and more children a fighting chance at survival. Dr. Semmelweis's investigation is the reason why there are signs in bathrooms across America today instructing employees to wash their hands.

Another breakthrough for the human life span was the introduction of antibiotic drugs like penicillin. Before the mid–twentieth century, simply trimming your rose garden could be a dangerous pastime. The discovery of antibiotics changed the way people lived their lives. As researchers beat back the bacteria, children, young adults, and women were all afforded a better

In mid-nineteenth-century Europe, one in two hundred women who had a baby did not survive the year after giving birth. It was basically taken for granted that after childbirth, a lot of women would catch something called puerperal fever, an infection of the reproductive organs that often leads to death.

chance of living longer. After penicillin was invented, the human life span jumped another ten years.

As more people aged, the ratio of what was killing us shifted from infectious diseases to the illnesses that accompany aging, like heart disease. By the 1950s, as people started living through infections and making it to their seventies in record numbers, cardiovascular diseases became the number one killer in the country for men and women. It is still the number one killer today, partially because the technological advances that increased life expectancy also led to modern conveniences that can undermine it: innovations like processed foods and devices that enable us to live sedentary lifestyles.

YEAR	FEMALE LIFE EXPECTANCY	MALE LIFE EXPECTANCY
1850	40	38
1900	51	48
1950	72	66
2000	80	75

HOW WE GOT HEART SMART

Until the middle of the twentieth century, the causes of heart disease were an utter mystery to the medical community. Most theories were based on speculation; there was not enough meaningful data to provide any real insight into how to prevent the disease that was killing so many Americans. Then in 1948, the National Heart Institute (a part of the NIH) launched a study that is still recognized as one of the most important medical accomplishments of the era: the Framingham Heart Study. More than five thousand men and women between the ages of thirty and sixty-two from the town of Framingham, Massachusetts were selected to participate in the longitudinal study, which assessed the health of these individuals every two years over the course of several decades. (The study is still running; in 1971, another group of more than five thousand participants enrolled, the children of the original group; in 2002, the grandchildren of the original group signed up.) Through tracking such a large sample of participants over time, researchers learned that the common denominators of heart disease included high blood pressure, smoking, and high cholesterol levels. A model for later longitudinal studies, the Framingham study gave doctors a way to identify possible heart disease candidates and to develop preventive measures. It was also the first major study that included women, an important qualification that we will discuss in more detail later.

At the time that the Framingham study began, treatments for age-related diseases were mainly aimed at easing the symptoms and making the patient comfortable, not actually curing the disease. But in the 1950s, doctors gained new ground in surgery. Insights gleaned from surgeons in World War II allowed surgical procedures to become more specialized. Doctors better understood and had better access to diagnostic tools, anesthesia, and blood transfusions. And the widespread availability of antibiotics made high-risk surgeries safer. Surgeons were more confident than ever, and their first order of business was the heart.

By the 1960s, heart transplant surgery was on its way. In the middle of the decade, a chimpanzee's heart was put into a man. In the late 1960s, the first human heart was transplanted. Revolutionary techniques like these, along with the rise of coronary care units, the treatment of high blood pressure, and the improved medical response to coronary disease, were part of the response to the threat that cardiovascular disease represented.

Between 1950 and the end of the twentieth century, although cardiovascular disease was still the number one killer of men and women, the number of such deaths dropped by half. At the same time, the study of cellular biology became more sophisticated. Better diagnostics and therapies stemmed from discoveries in biochemistry and physiology, such as the ultrasound and the CAT scan. Each development offered doctors and scientists the opportunity to utilize more sophisticated information about the human body. And as a result, human life expectancy got longer.

For human beings today, life may be getting longer still. We are starting to develop a much deeper understanding of the causes of many age-related diseases, how to treat their symptoms, and how to prevent some of these illnesses in the first place. Tests now exist that can screen for and accurately diagnose cancer and other chronic conditions before they progress to a fatal stage. Numerous medications are widely available to manage hypertension. And, just as we all now take it for granted that washing our hands is essential to limiting the spread of germs, there aren't many people alive today who aren't aware of the fact that smoking is bad for you.

It wasn't an accident or a random mutation of our genes that bought us these extra years of holidays, vacations, birthdays, and anniversaries. These bonus years are the result of developing a better understanding of the world we live in and the dangers that are present within it. They are the result of the efforts of countless individuals striving at universities, hospitals, and other learning centers, all asking "why" every day. The knowledge we have gained over the past 150 years has allowed us to casually reference our forties as "midlife"—and to be 100 percent correct.

PLEASE TRY THIS AT HOME

Many doctors say that compliance is the hardest part of keeping people healthy—they can tell us to treat our bodies better or prescribe medications, but they can't make us adhere to their advice.

Over the past several decades, public health officials have been working to raise awareness that habits like smoking, not getting enough exercise, and eating junk foods are unhealthy. Despite their efforts, nearly 20 percent of American adults still smoke. Nearly 70 percent of adults are obese or overweight. Less than 40 percent eat the recommended five servings of fruit or vegetables every day. These are all indicators that more people still need to hear, understand, and implement the message about making healthy choices.

The reality is that for a large percentage of Americans, choices are limited. Cost, access to healthy foods, and education are real issues that perpetuate the obesity crisis. Many low-income families subsist on inexpensive fast food or processed foods, both of which are high in unhealthy fats and salt as well as loaded with hidden sugars and preservatives. One of our greatest challenges as a society and as individuals is turning education into action, and helping individuals remove the obstacles that make unhealthy choices cheaper and more accessible than healthy choices.

The bottom line is that our lifestyles affect our risk factors for disease. If our life expectancy numbers don't continue to climb, and the next generation's life expectancy is lower than that of their parents, it is going to be the result of what they have learned at home.

THE COSTS OF THE SILVER TSUNAMI

The ability to live longer, to spend more years on this planet and more time with our families and loved ones, is an amazing opportunity. But just like with any opportunity we seize, there are also consequences.

For the *first time in history*, there are now more people over the age of sixty-five than under the age of five living on this planet. Researchers call this phenomenon the silver tsunami.

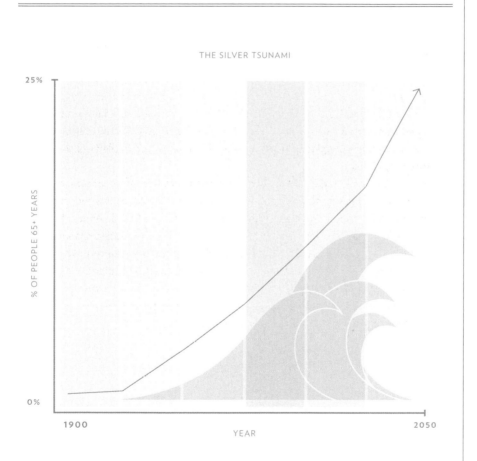

THE SILVER TSUNAMI

The silver tsunami makes aging a pertinent topic for every person, no matter when his or her birthday is. For most of human history, as we now know, people didn't live to be very old, so young children tipped the scale of the population—there were always more kids around than old people. But all that is about to change, and it is imperative that we consider what the impact will be on a social and economic level so that we can figure out how best to navigate the coming tide.

The first wave of the incoming tsunami will be made up of baby boomers, who were arguably the first generation of Americans to fully benefit from all the advances of modern medicine. The rest of us will be a part of subsequent waves. And as even more sophisticated medical technologies emerge, science will likely push the boundaries of life expectancy once again. All that longevity comes with a price. Healthy adults will retire later, which means that the people we vote into government will have to reconsider retirement ages and benefits, and people who run businesses will have to change the way they staff their companies. Traditional hierarchies may be overturned. Many older people may find themselves taking lower-level positions, reporting to people who are actually their junior in age but their senior in authority.

Sixty-six percent of caregivers are female. In the years ahead, more women than ever before will be called on to be caretakers for aging partners and relatives, a duty that will cost them in terms of both their physical and financial health.

Another thing to consider is that more living people will mean more sick people, and so along with increased numbers of older people we can expect that the cost of healthcare will increase as well. Not just for the government, but for private insurance companies, and for women everywhere. Why women? Because when it comes to caring for sick and aging spouses, parents, siblings, in-laws, and friends, it is women who are most likely to take on the responsibility of caregiving. Sixty-six percent of caregivers are female. In the years ahead, more women than ever before will be called on to be caretakers for aging partners and relatives, a duty that will cost them in terms of both their physical and financial health.

Women's caretaking in the United States is valued from $148 billion to $188 billion annually. Given that the ability to provide care for others reduces the hours women are able to work by around 40 percent, the total cost to an individual caretaker over time will be more than $300,000. It is known that long-term caregivers are likely to suffer the ill effects of stress, and are more likely to retire early with reduced pensions due to their reduced work hours.

The forecast for the silver tsunami is that it will affect all of us, and the impact will be personal and national, physical and emotional, economic and social and political, even environmental. Our planet will also pay the price as natural resources are stretched beyond limits to support so many more lives.

Did that forecast just throw a wrench into your excitement about a longer life span? Well, I urge you to take this information and add it to your own personal equation of how you'd like to proceed as an aging human, and specifically, as an aging woman—and then do your best not only to appreciate this extra time, but also to learn as much as you can about caring for yourself in the very best way that you can. And to encourage the loved ones that you are taking the journey with to care for themselves in the best ways they can. I believe that is our smartest strategy moving forward. It is the only strategy, really.

With the awareness that this wave is on its way, the most important thing we can do for our families and ourselves is to be strong enough to surf it.

CHAPTER 3

THE NEW SCIENCE OF GETTING OLD

How Aging Is Studied Today

————

S INCE WE HAVEN'T BEEN getting old for very long, it makes sense that the study of aging is also still in its infancy. The branch of the primary US agency dedicated to the research and understanding of aging was born around the same time that I was, in the early 1970s. It was only in 1974 that the National Institute on Aging (or NIA) was established, one of the twenty-seven institutes and centers that make up the National Institutes of Health (NIH), our nation's federally run medical research institute.

The NIA funds research aimed at understanding aging and improving healthy living as we age. It is also the primary federal funder of Alzheimer's research, which consumes much of its approximately one-billion-dollar budget for research grants (1/400th of the total NIH budget is dedicated to funding aging research). Some of that budget was spent in 2007 to kick-start the emerging field of geroscience; the NIH gave the Buck Institute, a nonprofit biomedical research center, $25 million to study aging and its link to chronic diseases.

When we were traveling across the country in November and December

2014 to learn about the science of aging, we met with a group of researchers who were in the midst of publishing a collaborative paper called "Geroscience" (you can look it up and read it online in the journal *Cell* if you like). Their paper made a compelling argument for a new, interdisciplinary approach to aging research.

Its authors are a mix of highly regarded scientists. Some, like Dr. Brian Kennedy at the Buck Institute, lead research initiatives for nonprofit organizations that study aging; others, like Dr. Elissa Epel at the University of California, teach and conduct research at major universities; and still others, like Dr. Felipe Sierra, the director of the Division of Aging Biology at the National Institute on Aging, hold leadership positions at government-funded entities. The publication of this paper marked the first time that scientists from the private sector, academia, government, and independent organizations came together to collectively investigate a new way to study aging. It also marked the first time that scientists from across a range of disciplines—from cell biologists to geneticists, endocrinologists, pharmacologists, and mathematicians—collaborated on this common goal.

And the authors of the paper suggested something revolutionary: that future research should approach the various diseases of aging as having a single shared root cause—aging itself.

THE OPPORTUNITY OF GEROSCIENCE

The field of geroscience aims to understand the relationship between aging and age-related diseases. The word root "gero" is derived from the name of the Greek god of aging, Geras. In classical Greek mythology, most of the gods were represented as young, strong, and beautiful humanlike creatures, but elderly Geras was depicted as shriveled and small. While he may not have been as buff as Zeus, Geras—which translates to mean "gift of honor," or "privilege of age," or "reward"—had other attributes to offer. Because as youth flees, we gain honor, courage, wisdom, experience, and other rewards.

The term "geroscience" feels apt as we study the new science of getting older, as Geras offers the perfect metaphor for the human relationship with aging: a fear of physical weakening coupled with the awareness that without years, without experience, the gifts of a life well spent cannot be fully realized. Today, geroscience is attempting to reconcile these two views by investigating how we can remain strong and vital as we age.

According to geroscientists, aging is the biggest single risk factor for chronic illnesses like cardiovascular disease, cancer, type 2 diabetes, osteoporosis, and neurodegeneration (including Alzheimer's). For decades, medicine has been studying the chronic diseases related to aging separately instead of collectively. By looking at heart disease as distinct from cancer as distinct from Alzheimer's, we miss a valuable opportunity to understand what they might all have in common. The radical question posited by the field of geroscience is: What if there were a different way to understand the process of aging and, in doing so, alter our rate of aging?

Dr. Gordon Lithgow, an expert in aging and genetics who is the principal investigator and director of the Buck Institute's Interdisciplinary Research Consortium on Geroscience, explained to us how the past revelations that have lengthened our life spans may mirror today's discoveries about aging. In the nineteenth century, life expectancy increased when scientists realized that many of the diseases that were killing people had a common cause.

The revolutionary question posited by the field of geroscience is: What if there were a different way to understand the process of aging and, in doing so, alter our rate of aging?

Tuberculosis, smallpox, and the flu may have looked differently and behaved differently, but they were all the result of coming into contact with tiny little organisms we couldn't yet see. The discovery of bacteria and viruses allowed scientists to develop effective treatments for the illnesses they cause.

The awareness sweeping science today is that the same principle may be true for the diseases of aging. Heart attacks, cancer, and diabetes all look different and behave differently, but if we can understand their common

WHAT'S THE DEAL WITH STEM CELLS?

One new area of medicine that's gotten a lot of attention recently is regenerative medicine, which uses stem cells to help heal and repair damaged and diseased organs. There's been a lot of excitement about the potential of stem cells to heal, as well as much controversy about how they are harvested.

There are two main classifications of naturally occurring stem cells: embryonic stem cells and adult stem cells. Embryonic stem cells have tremendous value in medical research because they have the ability to divide and become other types of cells. Embryonic stem cells come from human embryos, which contain cells that can differentiate into one of three kinds of primary cell layers (ectoderm, endoderm, and mesoderm) that have the potential to turn into any kind of cell in the body, from skin to muscle to nerve. But the practice of harvesting embryonic stem cells is controversial.

Then there are adult stem cells, also called somatic stem cells, which live in the tissues of our organs. We now know that adults have stem cells in our brains, bone marrow, blood vessels, skin, teeth, heart, gut, liver, ovaries, and testis. These powerful cells are always at the ready to heal and repair, and have the potential to morph into other types of cells needed by the organ. Stem cells can remain dormant for a long time until they are needed to make more cells, or until a disease or injury incentivizes them to spring into action. Some adult stem cells can also be activated following exercise (as if you needed another reason to get your body moving!). The challenge of treating disease with adult stem cells is that adults have few stem cells in our tissues, and even once those have been harvested and isolated in a lab, growing more of them isn't easy.

Enter induced pluripotent stem cells (iPSCs). The 2012 Nobel Prize was awarded for the discovery that mature cells—normal adult cells, like skin cells—can be reprogrammed to become immature, embryonic-like cells capable of developing into specialized cells. These time-machined cells can then be used for treatments throughout the body. Since this breakthrough, researchers around the globe have been creating iPSCs and encouraging them to divide to become skeletal cells, epithelial (tissue) cells, and cardiac cells to see if they can form new bones, skin, and hearts.

What's so exciting about iPSCs is that they provide a way around the embryo-harvesting issues and ensure a scalable supply of adult stem cells for research and treatment. Not only that, but these cells can be derived from specific people, and so it will be *your* cells that are potentially made into the new cardiac cells you need after a heart failure, or *your* cells that make the dopamine cells you need to replace those lost in Parkinson's disease.

Over the coming years, we will be hearing more about the future of regenerative medicine and transplants. Scientists are very optimistic about the therapeutic potential of stem cells.

cause, we may be able not only to live longer, but also to age with more of our health intact.

Think about that for a moment. If aging is the common cause of those illnesses, and we can increase our understanding of aging at the cellular level, we may be able to live with strength and health until we die quietly in our sleep. Our bodies will weaken naturally, and aging puts us at risk for a host of diseases, but a risk factor is not a diagnosis. It's a call to action, to arms, and to attention. Knowing the risks can empower us to become the architects of our own strength and resilience.

HOW AGING IS STUDIED

We are aging in a time when science is committed to and compelled by the question "What is aging?" The science of understanding aging takes place in laboratories and meeting rooms, at desks and with the help of technology. Study participants may dutifully take medications, try different ways of eating, of moving, of sleeping, or just give up their privacy and answer loads of questions, all so that we can better understand how aging affects our bodies. The test groups for these studies range in sample size from a hundred people to hundreds of thousands of people. When we see the results of the latest research plastered all over social media or mentioned in a morning news show or even announced in the headlines of newspapers, it's important to keep in mind that every study varies in terms of how it collects its data. Data can be influenced by how many participants are included, and also by how well a study is designed and the elements for which its researchers are controlling—factors ranging from time and temperature to gender and age may affect a study's accuracy.

One type of research method is the observational study. Observational studies assess how the choices people make affect their wellness. Some of these studies, like the Framingham heart study, are longitudinal — they track participants over a sustained period of time. Longitudinal studies can last for decades, and they have been very useful in helping us understand aging.

Observational studies can also examine a cross-section of people and compare them with one another to see how their choices have influenced their health. For example, one cross-sectional study compared older people with similar health profiles to determine the effects of vitamin D deficiency (answer: it can put you in a lousy mood as well as impair your ability to think clearly).

Some studies are less about observing, and are more about getting involved. An interventional study gathers groups of people to study for comparison. Researchers ask one group to implement a specific behavior in order to observe its impact on health, and use the other group as a control for comparison. For instance, when researchers wanted to understand how the intensity levels of physical activity would affect memory, they organized sixty-two healthy older people into three groups. Over the course of six months, one group performed medium-intensity workouts, one group performed low-intensity workouts, and the other group did not work out at all. The study found that any exercise had great benefits for memory and brain volume, with little difference between the low- and medium-intensity groups.

Our bodies will weaken naturally, and aging puts as at risk for a host of diseases, but a risk factor is not a diagnosis. It's a call to action, to arms, and to attention. Knowing the risks can empower us to become the architects of our own strength and resilience.

When researchers study aging, they don't only observe humans, they also turn to human cells, animal cells, bacteria, mold, and fungi. They experiment on lab animals (and sometimes wild animals), from sponges and worms to mice and monkeys to naked mole rats. By examining and manipulating genes from other animals, scientists can gain insights into the human aging process. Although worms look nothing like human beings on the outside, internally, like flies, they have many genes and biological mechanisms in common with us. And just like people, worms and flies age, albeit at a much faster rate. Within a year's time, many rounds of testing and learning can take place on worms and flies, whereas watching humans age in observational studies takes a human lifetime—more time than one scientist has on her hands.

Comparative biology can also yield important insights. Naked mole rats, which are somewhat terrifying-looking mouse-sized rodents, live seventeen to twenty-eight years, while common mice and rats live only three to five years. Why the disparity? By comparing similar animals with very different life spans, scientists may be able to identify the specific genes that make some mammals live longer than others.

We know from being avid consumers of media that certain things are "good for us," but we may not fully understand *why*. So let's do our best to learn these things. Let's try to better understand how our choices influence our health at the cellular level, and how the changes in our cells are what affect our health as a whole, and in particular, our health as we age.

If one direction of learning shows promise when it comes to increasing the length of life—and the length of a healthy life—scientists will apply and reapply methodologies that are more and more complex, ultimately testing medications or treatments on humans to determine their safety. At each step, they must evaluate the efficacy of their methods. Just because manipulating a certain gene or pathway or hormone makes a fly live longer doesn't mean it will do the same for a mouse, let alone a human.

And just because it works for a man doesn't mean it will work for a woman, as we will soon discuss.

IS IT POSSIBLE TO GROW OLD WITH HEALTH?

Life expectancy doesn't tell us how long a being might live under the best of conditions. It tells us how long a being might live while taking into account the reality of its environment. We can give science a lot of credit for doubling our life span, but we can't give doctors all the responsibility for keeping us healthy.

Your genes create the basis for your health. The environment you live in and the lifestyle choices you make every day have a massive impact on how you age, what makes you sick, and how your body heals. As we grow older, weakening is inevitable and becoming strong is a choice. Diseases of old age are not necessarily a given for any one of us. I think it's always more challenging to make new choices or try out new behaviors if you don't understand the "why" behind them. Advice like "eat healthfully" or "exercise every day" doesn't really mean anything to me in a vacuum. That's why the information in this book is so important to me. Without context, how can any of us be expected to understand why we should eat more vegetables or why strength training is a big deal for women? We know from being avid consumers of media that certain things are "good for us," but we may not fully understand *why*.

So let's do our best to learn these things. Let's try to better understand how our choices influence our health at the cellular level, and how the changes in our cells are what affect our health as a whole, and in particular, our health as we age. We can become better advocates for our own health. It all starts with learning the facts.

CHAPTER 4

SEX, DRUGS, AND BIKINI MEDICINE

*How Being Female Affects Your
Health and Your Healthcare*

F ACT: **WOMEN LIVE LONGER** than men. A baby girl born in the United States in 2010 had a life expectancy of eighty-one; for the baby boy next door, that number was seventy-six. That's a five-year gap, enough to make a person really curious about why this might be the case, especially when you consider that this is true around the world, too. Country by country, life expectancies vary (due in part to variables like availability of clean water, access to healthcare, and stability of the region), but the world over, the women's life expectancies are always greater than men's.

In the United States, over the course of all that longevity, women use more healthcare services and take more prescription drugs than men do. Researchers at the Mayo Clinic made headlines a few years back when they announced that nearly 70 percent of Americans take at least one prescription drug daily. They also reported that, as a whole, women and older adults receive the most drug prescriptions. As people get older, they are prescribed more pills to take, not fewer, and the quality and accuracy of those medications has a direct impact on our health. The more orange bottles lined up in your bathroom cabinet, the more important it becomes that you are taking the right medications, in the right doses, at the right times.

Health and healthcare are inextricably bound together. Every time you go to the doctor's office, every time you pop a pill, you are relying not only on your physician and your pharmacist, but also on medical schools, on drug companies, on research labs, on individual scientists—and their assumptions about women, and their awareness of the latest research about women's health and women's bodies.

A lot of people hate going to the doctor, and I get it. Hanging out in a waiting room on your lunch hour or having blood drawn when you're running late for an appointment is not exactly fun. But I take going to the doctor very seriously. When I'm sick, I make an appointment. And when I'm healthy, I make appointments, so that I can avoid getting sick for as long as possible. I want to understand where my health is now so that I have a framework for comparison for later. I want to use medicine as a preventative tool for my health as I age.

And researchers are discovering that this habit of mine might actually be tied to female longevity. Women are more likely to visit the doctor than men and this may help us to live longer. So do the other healthy choices that, as a group, women make more than men do, like not smoking. Fewer women than

> Every time you go to the doctor's office, every time you pop a pill, you are relying not only on your physician and your pharmacist, but also on medical schools, on drug companies, on research labs, on individual scientists—and their assumptions about women.

men smoke, which cuts our risk profile for numerous diseases. Men also drink more than women do. Women are more careful about their nutrition, and taking care of food needs helps bolster strength and longevity. And women are less likely to take risks. Fewer risks equals fewer injuries, which equals greater health: unintentional injury is number three on the Centers for Disease Control and Prevention's mortality charts for men, and number six for women. Women also value friendship, love, and connection. We are social beings who invest in our families and in our communities and in our relationships. All these choices contribute to a longer life.

But female longevity isn't just about our choices. Among primates like chimpanzees, females live longer too, and monkeys don't make doctor's appointments. So why do females enjoy a longer life span? Some scientists are looking for answers that are rooted deep within the genetic coding of our cells. The cells of men and the cells of women are not the same, and what makes your cells unique affects everything about you—including how long you live.

THE OLDEST WOMAN IN THE WORLD

Today, life expectancy is twice what it used to be, but it may still be forty years shy of the maximum human life span, which most scientists believe to be about 120 years old.

They base that opinion on people like Jeanne Calment, the oldest woman who ever lived. Jeanne was born in France in 1875 and passed away in 1997. When she was a year old, Alexander Graham Bell invented the telephone. When she was thirteen, she met Vincent van Gogh. When she was eighteen, the Wright brothers flew for the first time. She lived through two world wars, saw infections thwarted by medicine, and witnessed the development of the Internet and contemporary medical technologies.

When she was ninety, a lawyer who was not yet fifty offered to pay her every year if he could take over her home when she died. She agreed. He died at seventy-seven, and Jeanne kept on going. She lived by herself until she passed away at the age of 122.

WHAT DOES IT MEAN TO BE BIOLOGICALLY FEMALE?

You are a lady with millions of microscopic lady cells. Your female cells are special—the genetic information contained in each and every one of them is what makes your body biologically distinct from that of a man. Your female cells have distinct characteristics, and so do the organs they make up, because female organs are sized and shaped differently than male organs.

For example, the female heart has a distinct architecture. Our heart is smaller than a man's, with thin vessels arranged in a lacy pattern instead of the thicker tubes that connect a male heart to his cardiovascular system. Cholesterol plaques can settle throughout a woman's arteries, instead of in a more obvious clump, as tends to happen in men's hearts, making it more challenging to detect heart disease in women. That is why women experience heart attack symptoms differently than men do, and why they need different care at the hospital during and afterwards. Women are far more likely than men to receive an incorrect diagnosis about heart attack symptoms despite the fact that more women than men die of cardiovascular disease every year in the U.S.

Our hearts also beat in a distinct rhythm; any disruption of that rhythm results in an irregular heartbeat, called an arrhythmia, which can range in severity from mildly disruptive to life threatening. In addition, some medications and defibrillators—which have primarily been tested on men—have been shown to cause potentially fatal complications in women.

Your female cells are special—the genetic information contained in each and every one of them is what makes your body biologically distinct from that of a man. Your female cells have distinct characteristics, and so do the organs they make up, because female organs are sized and shaped differently than male organs.

Women's unique biology also translates into unique risk factors for disease. For example, women are more likely than men to develop depression, eating disorders, and anxiety disorders like post-traumatic stress disorder. Mental health issues are, in turn, a risk factor for a variety of other diseases. A twelve-year study of more than 10,000 women in Australia aged 47–52 showed that middle-aged women who were depressed have twice the risk of having a stroke compared to women who are not depressed.

But until the past few decades, nobody was talking about female cells, and it's taken some time for the medical and scientific communities to realize just how important it is to consider the sex of our cells when it comes to healthcare. Today, with an improved understanding of how sex-specific biology affects research, diagnostics, and medical treatment, women are getting better care.

HOW DO CELLS GET A SEX?

Healthy people have forty-six chromosomes in each of their cells. Two of your forty-six chromosomes determine sex. The rest, the autosomes, determine pretty much everything else about you. Sex chromosomes only come in two varieties: X and Y. Women are XX and men are XY. The X chromosome is much larger than the Y chromosome, because it is a powerhouse: not only does it determine sex, it also contains additional genes and thus additional information. The Y chromosome is smaller and carries less information.

Your sex is determined by your father. When a sperm, which has twenty-three chromosomes, twenty-two autosomes, and one sex chromosome, collides with an egg, which has twenty-three chromosomes, twenty-two autosomes and one sex chromosome, fertilization takes place. The result is a complete human cell—called a zygote.

1 Female Chromosome

2 Male Chromosome

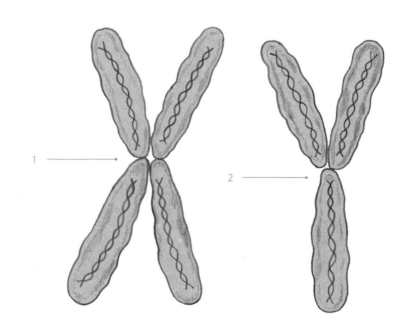

Since all of a woman's cells have an XX, her eggs can only have an X. Since men have an XY, their sperm can contain either an X or a Y. It's a toss-up. Will a fertilized egg become a female or a male? It's all about the sex chromosome carried by the sperm.

If you are a woman, the sperm that made you carried an X chromosome. As the female zygote divided, all the cells it made were female, and thus, so are you. A zygote divides and replicates and divides and replicates and divides and replicates. And after all that replication, you've got a cluster of female embryonic stem cells that will eventually keep dividing to become heart cells and liver cells and skin cells and blood cells and brain cells. And all those resultant organs will be female organs made up of female cells.

The X chromosome is much larger than the Y chromosome, because it is a powerhouse: not only does it determine sex, it also contains additional genes and thus additional information. The Y chromosome is smaller and carries less information.

The genes located along the X chromosome are called sex-linked genes; some of them are coded specifically for female anatomical traits, but other genes are responsible for more than three hundred genetic diseases, known as X-linked disorders. The fact that we have two X chromosomes in each of our cells may actually be one reason why women live longer than men—if one of our Xs contains a faulty gene, we have another X to step in. One example of this is red-green color blindness, a common X-linked disorder. More men than women are color-blind, because if the gene for color is disturbed on their X chromosome, they don't have a second one to use for backup. Having two Xs is like having an extra black dress in the closet—or even better, having a few extra years of life in which to wear it.

Just another reason to be grateful that you're a lady.

For centuries, the medical community viewed women as simply smaller versions of men with different reproductive organs—organs that were thought to make us mentally unstable. You can go to a library and view 1,500 years' worth of medical journals—from classical Greek texts to Victorian literature—that record "hysteria" as a common diagnosis for women. The word "hysteria" is actually derived from the word *hysterikos*, Greek for "of the womb."

If a woman seemed psychologically stressed, or she started forgetting things, or she seemed emotionally volatile, like she needed too much attention, a physician might diagnose her with hysteria. Remedies ranged from sniffing smelling salts to having sex. A single woman would be advised to marry. Married women were advised to sleep with their husbands or to go horseback riding. Alternatively, in a not uncommon practice, doctors and midwives would use their fingers to stimulate women to orgasm, or "hysterical paroxysms"—basically a happy ending at the doctor's office.

As recently as the late nineteenth century, nearly three-quarters of the female population were deemed to be "out of health," including women with "a tendency to cause trouble." Since their issues generally couldn't be resolved in just one visit, these women represented America's largest therapy services market.

In 1980, "hysterical neurosis" was finally dropped as an official diagnosis by the American Psychiatric Association.

THE FALL OF BIKINI MEDICINE

If you ask a female doctor in her fifties about the model for human anatomy she studied in med school, she will tell you about the 70 kg (approximately 160-pound) man that was the example for her medical training—and his female companion, the 60 kg (approximately 130-pound) man with boobs and a uterus. Medical students were not trained to treat the male and female bodies as all that different, except when it came to the reproductive organs—the areas covered by a bikini. Hence the nickname for the medical education of the era: bikini medicine.

Bikini medicine was the product of centuries of misunderstanding about female anatomy that began with all that hysteria about hysteria. The fall of this antiquated model and the rise of our more accurate understanding of female biology is tied in large part to the strides made by a generation of thinkers and speakers and brave, determined individuals who demanded change as a part of the women's rights movement. The modern view of women's health we take for granted today only began in the 1960s, as women across the country sought information about their biology and demanded better-quality medicine, including better reproductive care and uninhibited reproductive rights. They made it clear to lawmakers and medical professionals that women's legal rights and medical welfare were inextricably linked.

The fall of bikini medicine and the rise of our more accurate understanding of female biology is tied in large part to the strides made by a generation of brave, determined individuals who demanded change as a part of the women's rights movement.

Here's one example: the year that I was born, 1972, was the same year that the federal government allowed unmarried women legal access to take birth control. Think about that. If you were a single woman in 1970, it was actually *illegal* for you to take charge of your reproductive system. In 1989, when I was seventeen, Congress allocated funds specifically for the study of women's health. By the 1990s, 30 percent of ob-gyn specialists were women, up from just 7% in the 1960s. Their efforts, along with the efforts of countless other doctors and scientists, increased the focus and attention on women's health, including the health of older women, setting the stage for the healthcare we will all receive in the years to follow.

1920:

Women get the right to vote

1971:

Ten percent of medical students are women.

1916:

Margaret Sanger opens the first birth-control clinic in Brooklyn. Ten days after it opens, the police shut it down and put her in prison. Contraception is illegal.

1963:

Congress passes the Equal Pay Act.

1960s:

The women's health movement begins.

1916:

Planned Parenthood is founded.

1960:

The birth control pill is approved by the Food and Drug Administration (FDA).

1967:

The National Organization for Women (NOW) is launched.

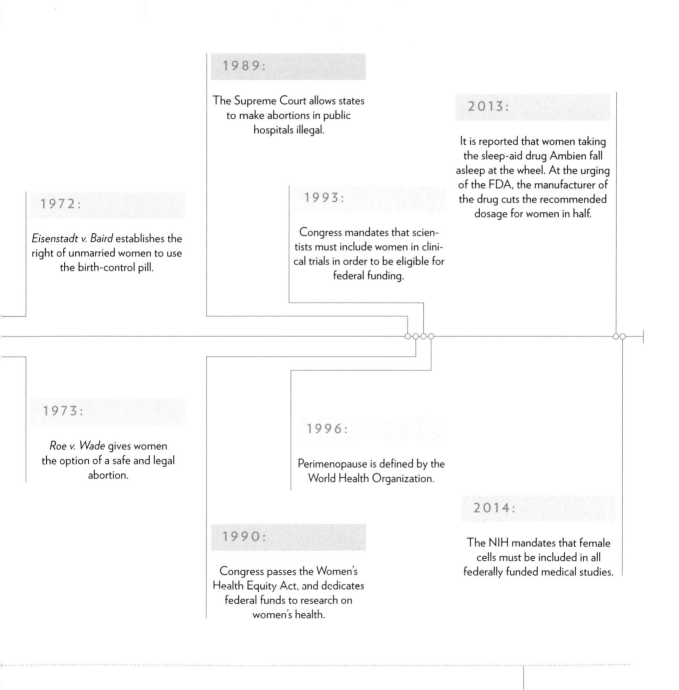

1989:

The Supreme Court allows states to make abortions in public hospitals illegal.

2013:

It is reported that women taking the sleep-aid drug Ambien fall asleep at the wheel. At the urging of the FDA, the manufacturer of the drug cuts the recommended dosage for women in half.

1972:

Eisenstadt v. Baird establishes the right of unmarried women to use the birth-control pill.

1993:

Congress mandates that scientists must include women in clinical trials in order to be eligible for federal funding.

1973:

Roe v. Wade gives women the option of a safe and legal abortion.

1996:

Perimenopause is defined by the World Health Organization.

1990:

Congress passes the Women's Health Equity Act, and dedicates federal funds to research on women's health.

2014:

The NIH mandates that female cells must be included in all federally funded medical studies.

HOW YOUR SEX AFFECTS YOUR DRUGS

During our visit to the NIH, we spent valuable time with Dr. Janine Clayton, the director of the NIH's Office of Research on Women's Health (ORWH). The ORWH has been around for only about fifteen years. Its mission is to promote women's health initiatives in the medical community, educate the public about issues related to women's health, and fund programs that explore the role of sex and gender differences in medicine. Dr. Clayton and her team have been working for years to encourage her colleagues at the NIH to prioritize women's health issues and needs in their research.

Dr. Clayton is an intelligent, educated, personable, and powerful woman. So of course we asked her questions about sex and drugs.

Your female sex affects the efficacy and the potency of the drugs you take. This is especially important because as a nation, we take a lot of drugs. But while women are being prescribed more medications than ever before, not all of those medications are properly tested for use by women. And it has been proven again and again that drugs don't affect men and women the same way.

When we swallow a pill or get an injection of a vaccine, the medicine travels throughout the body via the bloodstream and is distributed to our tissues and organs. Medicines affect women differently than they affect men for several reasons:

WE HAVE DIFFERENT ORGANS:
A female liver metabolizes drugs differently than a male liver.

WE HAVE DIFFERENT BODY WEIGHTS:
Men are usually bigger and heavier, and have bigger organs, requiring higher dosages of drugs than a smaller woman does.

WE HAVE DIFFERENT BODY COMPOSITION:
Women store more body fat than men do, and some medications are attracted to fat tissues. When a woman takes those drugs, they linger longer in our bodies, and their effects linger too.

WE HAVE FEMALE HORMONES:

Our hormones influence how our bodies process medications. Factors like oral contraceptives, menopause, and postmenopausal hormone treatment could also affect how we respond to drugs.

Painkillers and anesthetic drugs are absorbed and metabolized in a unique way by women, who have a 30 percent higher sensitivity to neuromuscular blockers and in turn need smaller doses than men. Research has shown that males and females do not respond in similar ways to opioids like OxyContin, Percocet, and Vicodin. Some medications, like Valium, exit our bodies faster than they do men's bodies. Others linger longer. In animal studies, males and females also react to withdrawal in different ways. It is crucial for us to be aware of these differences, because as a nation we are currently experiencing what the NIH has termed an "epidemic" of women overdosing on painkillers, with a dramatic rise in the number of women dying every year. And the highest risk of death from an overdose of prescription painkillers isn't found among the young—it is found among women between the ages of forty-five to fifty-four.

Even when it comes to animal testing for drugs being developed to treat illnesses that predominantly affect women, sex isn't always taken into consideration. That goes for human female subjects, too.

But all this information is relatively new, and that's because for a long time, pharmaceutical companies tested their drugs only on male cells, male animals, and male humans. This practice has led to a mountain of data that isn't very accurate when it comes to prescribing drugs for women. Many of the studies we rely on today faithfully and obsessively record variables like time and temperature, but overlook the small detail of sex. Even when it comes to animal testing for drugs being developed to treat illnesses that predominantly affect women, sex isn't always taken into consideration. That goes for human female subjects, too. Since hormones fluctuate over the course of a month, tests that use females can be a lot more complicated to analyze than tests that use males. Without taking hormonal shifts into account, it is impossible to determine how treatments might affect a woman over the course of a month. Pregnancy is a concern in medical testing as well. In the 1960s, thousands of pregnant women who participated in a study and took the drug thalidomide gave birth to babies with serious defects. That tragedy has had a lingering effect on the scientific community, making researchers wary of including women in drug trials at all. By not creating safer trials and including women, however, we've also lost the chance to gather important information.

Luckily, things are changing.

Since 2014, the NIH has required applicants for federally funded research grants to address how sex relates to the way experiments are designed and analyzed. This research is critically needed, because as you've read in this chapter, medications affect women uniquely—even everyday ones, like the flu vaccine. A woman requires half as much flu vaccine as a man to potentially produce the same amount of antibodies.

There are more than thirty million women between the ages of thirty-five and fifty living in the United States. Whatever health challenges you might be going through, you are not alone. We're millions strong.

Women need appropriate doses of everyday and life-saving medicines that have been developed to be effective for our bodies. We need research that supports our sex, our cells, and our lives. The more knowledge we have of our female biology, the more we can advocate for quality care for ourselves as we age.

Over the course of my lifetime, women's healthcare has improved dramatically. When I was a girl, the medical community was evolving in ways that would profoundly affect my life as a woman. I may not have been aware of the social changes that were swirling around me—and I certainly wouldn't have understood that the advances being made in women's rights were so closely tied to advances in women's healthcare. Now I understand that, ladies, we have been living through history.

As we are writing this book, there are more than thirty million women between the ages of thirty-five and fifty living in the United States. Whatever health challenges you might be going through, you are not alone. We're millions strong. We're standing in the middle of a conversation that has been going on for hundreds of years, with hard-won rights and knowledge bestowed on us by previous generations of women (and men).

And the changes are still coming.

WANT THE FACTS? ASK.

Whether you're choosing a phone plan, buying new clothes, or ordering from a dinner menu—chances are you probably ask a lot of questions before making a decision. How many minutes will I get? Do these jeans come in petite? Is the pasta homemade?

Asking questions helps you make sure you're getting what you want and need. So do you bring those same sleuthing skills to your doctors' visits? Unlimited texting, the perfect pair of pants, and an amazing meal are totally worth the time it takes you to assess your options—and so is your healthcare.

When we were at the NIH we met with its director, Dr. Francis Collins, and asked him what he thought the public needed to know when it comes to healthcare. He said that the most crucial thing we can all do is pay attention. Aging research (and other medical knowledge) is constantly evolving. We can't rely on medicine published twenty years ago for our treatment today.

The only way to do this is to ask your doctor lots of questions, and to keep on asking if you still aren't sure of the answer. We talked to doctors who said they were amazed by the LACK of questions they receive from patients. They told us their patients don't always ask about the medications they're prescribed—what they're for, what they do, what the benefits and risks and side effects may be. They all encouraged us to encourage you to ask more questions.

Everyone deserves to have a doctor who listens to their questions and takes the time to answer them. Who will discuss alternatives and options. And who respects and takes seriously their symptoms.

Dr. Seth Uretsky, a cardiovascular specialist and medical director of cardiovascular imaging at Morristown Medical Center, often sees patients who feel their symptoms have been overlooked by physicians. He believes that it's important to trust and feel comfortable with your doctor, and that finding a doctor who listens, gives you time, and explains his or her thought process is crucial. And the data underscores his point: when patients feel understood by their physicians, outcomes are better.

If your doctor doesn't seem to have the time to listen to your concerns, or if you feel that he or she doesn't take your questions or symptoms seriously, it's time to find a new doctor.

PART 2

STEEP GRADES, SHARP CURVES

The Biology of Aging

FIFTY WAS A SHOCK, BECAUSE IT WAS THE END
OF THE CENTER PERIOD OF LIFE. BUT ONCE I GOT OVER THAT,
SIXTY WAS GREAT. SEVENTY WAS GREAT. AND I LOVED,
I SERIOUSLY LOVED AGING.

— GLORIA STEINEM

CHAPTER 5

TIME IS RELATIVE

The Biological Impact of Genes,
Choices, and Attitudes

W HEN I WAS A child, I loved to spend as much time as I could with my grandparents. My grandmother was my hero. She was seven years older than my grandfather, and she was a powerhouse.

My grandmother was my ideal of strength and ability. She basically maintained a full working farm at her house, raising her own livestock and tending a thriving garden. She didn't drive, so when she needed more feed for her chickens and rabbits she would set out on foot. She would walk to the feed store, about a mile away, and carry two ten-pound sacks of feed, one in each hand, on the walk back, even in the heat of the summer. If she wanted to haul more, she would bring a wagon. As a child, I remember asking her why she carried such heavy bags so far in the scorching heat. She would say, "Because I like it, and because it keeps me strong." Her answer was so awesome to me, and I think of it every time I am pushing myself to go farther, to work harder, to try a little bit more.

When that little voice in my head says "Stop! Why? Enough!" I say, "Because I like it. Because it keeps me strong." That is the legacy my grandmother left me, and I thank her for it each and every day.

My grandfather, although he lived in the same house with my grandmother, had a very different lifestyle than she did. His job required him to sit in an office, he smoked and he chewed tobacco, and he enjoyed red meat more than his vegetables. When I was eight years old, he had a heart attack and passed away. He was only sixty-three. Meanwhile, my grandmother lived to be nearly ninety. She stayed vital into her seventies. Hauling feed, along with all of her other regular chores, kept her strong and resilient.

Your birthday does not determine how long or well you will live. You may share the same birth year as someone else, or even the same birth date and birth mother, but the way that *you* live will affect the way that *you* age.

But even hard work cannot always protect us from illness. At seventy-three, my grandmother was diagnosed with breast cancer, which is now considered to be an age-related disease. She also developed an irregular heartbeat. So she underwent two major surgeries to save her life—she had a mastectomy and she had a pacemaker implanted. During her recovery she came to stay with us, and I spent that summer making meals, helping her bathe, and reading to her as she drifted off to sleep. She had a bell that she would ring in the morning, and I would come and take her to use the bathroom. For the first time in her life she was relieved of her strenuous chores and her day-to-day responsibilities—and she liked having her family close by, helping her, caring for her.

When she got better, she went back to her active lifestyle, gardening and working around the house. But she was getting older and getting tired. She didn't have the same energy that she'd had abundantly before her battle with cancer. And so, after a while, she decided that she had worked hard enough. She didn't want to take care of the animals and the garden anymore. She wanted a less strenuous life—and so my mom and dad asked her to move in with us.

At first she remained active, walking around the house and hanging out in the kitchen. But as time went by she began to rely more and more on our caretaking, and I watched her go from functioning mostly on her own to relying on me to hold her up as she scuffled with her walker. She went from sitting with the family in the living room to spending most of her day lying down and asking everyone to come visit her in her bedroom. She stopped eating her meals at the table and requested to have them in bed. As her effort declined, so did her health. I could understand how at her age she might want a rest from her backbreaking chores, but it was hard and confusing for me to watch her change. She was never the same again.

I was just a kid, but her changes imprinted on me how fragile even the strongest people could become. I saw how important it is to work as hard as we can, for as long as we can, if we want to age with strength.

WHERE DID YOU LEARN ABOUT AGING?

I witnessed different kinds of aging when I was a kid, and it taught me a lot about what I might expect for myself. Where did your first ideas about aging come from? Many of us learned about what aging looks like by watching our family members grow older. Were your grandparents healthy and active, or did they always seem old and infirm? How about your parents? The attitudes we absorb about what it means to age and what life is like for the aged will affect our own aging process.

Your birthday does not determine how long or well you will live. You may share the same birth year as someone else, or even the same birth date and birth mother, but the way that *you* live will affect the way that *you* age. While there are broad age-related changes that apply to everyone, for an individual, biological age (how healthy you are on the inside) is a more important indicator of health than chronological age (the number of years you've been on this planet). The aging of our cells is the true measure of how old we are.

Many different factors influence the way you age. There's the genetic component, of course, which we are reminded of every time we give our family medical history at the doctor's office. But genes aren't everything. While some research ties longevity to the genes we inherit from our parents, our choices, our environment, and our attitude have just as powerful an impact on how healthfully and long we live. How old we truly are—in a biological sense—is a combination of these factors.

My grandmother's habit of hauling chicken feed on hot summer days at the age of seventy is a good example of how the number of years we have lived and the level of health and independence we enjoy do not always correlate. Each of my grandparents aged in a way that was purely *of them*, not only genetically but also an extension of their personalities and how they lived their lives. When I was growing up, I was aware that my grandmother was seven years older than my grandfather. But now I understand that biologically, deep within her cells, she was probably a lot younger than her chronological age.

Nobody can resist the pull of time forever, but the longer we are able to find the energy and the reserves to keep pushing ourselves, to believe that aging with strength is possible, the stronger and more self-reliant we are likely to remain.

AGING IS PERSONAL

Nature gives you the genes you have at birth. As an adult, the environment you live in and the lifestyle choices you make help determine how you age and how you feel. Since where you live and how you live is personal to you as an individual, so is the way that you age. That's part of why the study of aging is so complex: there is no one-size-fits-all answer. You start out as a cluster of cells that develops in response to the genes that are coded deep within your DNA, which are also known as your genotype. Your genotype is what you inherited from both of your parents, basically the blueprint that nature has used to create and build you. When infants are born, they are immediately weighed and measured. This is a useful marker for health, because babies can be compared by height and weight. But by the time we are toddlers, our environments have already begun to affect our development as much as our genes.

Even in our earliest years, our environments influence our health. Growing up in a household that is very stressful, for example, will affect your genes differently than if you are raised in a home that is calm and secure. Alexandra Crosswell, a research scientist at the University of California, San Francisco, whom we saw give a rousing lecture about her work, has been studying a group of breast-cancer patients to investigate the ways exposure to stress during our childhood years can affect our health as adults. She found that participants who had experienced abuse, neglect, or chaos within their home as children showed biomarkers for inflammation later in life that are linked to negative outcomes for health and well-being.

The DNA each one of us receives from our parents is influenced by the lifestyle of not only our parents, but also our grandparents and our great-grandparents and our great-great-grandparents. The study of this biological phenomenon is called epigenetics.

Life has a real effect on our health. Stress and trauma and smoking have an effect; love and safety and eating green vegetables have an effect. But these factors don't just have a short-term impact on our well-being; as Dr. Crosswell's study suggests, they can also have a very powerful, long-term influence. How is this possible? Because your lifestyle—the experiences you have, the places you live, the choices you make—can actually alter the way your genes are expressed.

Some genes are like light switches that can be flipped on or off by experience. Good nutrition, fitness, and low levels of stress may help flip off some of your genetic predispositions to disease, while smoking, eating poorly, and being sedentary may flip on your genes for disease. And the genes that have become altered as a result of your lifestyle can actually be passed down to the next generation. That means that the DNA each one of us receives from our parents is influenced by the lifestyle of not only our parents, but also our grandparents and our great-grandparents and our great-great-grandparents. The study of this biological phenomenon is called epigenetics. A relatively new branch of the field of genetics, epigenetics suggests that the effects of life on our genes can be passed down like family heirlooms.

My grandmother's active nature affected my attitude about aging. But the effects of her choices on my health may be even deeper than attitude. If she was especially active when she was young, then those genes could have been passed on to my mother, and to me. If I wind up enjoying a long life, with great health and strength as I get older, will that be because my grandmother was strong and fit in her youth? Or will it be because I am actively learning about health and my body, and doing my best to take care of my physical and emotional self? If I age well, will it be because of me? Or because of my genes? In the end, it will be both. My genotype has been affected by the life I've lived. This mix of nature and nurture has created what is known as my phenotype—the "observable" characteristics that are the result of the interaction between genes and environment.

My phenotype, and yours, includes physical traits like hair and eye color, as well as behavioral and personality traits like whether or not you gain weight easily or how anxious or laid-back you tend to be. We all have a pheno-

As we age, our bodies will undergo a series of shifts and changes, detectable at the cellular level and visible at the surface level. And these changes will not affect you and your friends or your siblings in the same way.

type, but each phenotype is distinct: nobody else in the world has your phenotype. Even if you're an identical twin—which means you share a genotype with your sibling—your phenotype is still unique. That's because it's impossible to live the same exact life, making the same exact choices and having the exact same experiences as anyone else. As we age, our phenotypes have a significant impact on our health and on our likelihood of developing the diseases of aging. Genotype is nature. Phenotype is nature and nurture mixed together.

As we age, our bodies will undergo a series of shifts and changes, detectable at the cellular level and visible at the surface level. And these changes will not affect you and your friends or your siblings in the same way.

AGING THROUGH THE DECADES

Your body has always been changing, and that's because it is an amazing machine that can grow, heal, reproduce, and, yes, age. You may currently be at an age at which you have witnessed enough changes to finally accept that over the rest of your lifetime, your body will shift a few more times. The rate at which we age and the shifts that accompany aging are unique for every person. We will all age, but we will not all experience aging the same way. Yet there are, generally speaking, some changes that will affect us all.

The changes of aging began when you were fairly young, because in organisms from worms to humans, the onset of aging happens around the time that sexual maturity is reached. That means that right when your boobs

Right when your boobs were at their perkiest, when you weren't thinking about getting older except maybe to turn twenty-one so you could get into bars and clubs legally, the aging process was starting to take hold deep within your body.

were at their perkiest, when you weren't thinking about getting older except maybe to turn twenty-one so you could get into bars and clubs legally, the aging process was starting to take hold deep within your body.

The changes first began to show up in your tissues and organs around the time you got that legal ID. The tissue of your lungs, those gorgeous balloons that give us breath, began to lose their elasticity and the muscles that surround and support your rib cage started to shrink. That means that with every breath you have taken since you were of legal drinking age, you inhaled just a little bit less oxygen. Of course, you probably never noticed. Perhaps because, according to some research, cognitive skills also begin to slip after age twenty-four.

In your thirties, the changes taking place inside your body are ones that you begin to see and feel. For instance, you begin to lose mass in your muscles, reducing their function, partially because you have fewer growth hormones and less testosterone than you did a decade prior. Around the same age, the number of cells in your kidneys decreases and the organs actually become smaller as a result, to the point where they filter blood less efficiently. At around age thirty you will have reached peak bone mass, making it extremely important to build up your bone mass before the age of thirty-five through nutrition and fitness. (More on that in Chapter 9.)

As your forties come into focus, you might find it tougher to read the menu in a dimly lit restaurant. That's because the lens of the eye thickens with age. Once you've secured yourself a stylish pair of reading glasses you may also notice some other signs of aging: gray hair due to decreased melanin production in your hair follicle cells, and fine lines, due to decreased collagen production in your skin cells. Your back may begin to ache, a result of tight muscles, herniated discs, or just too much sitting. And if you have that extra glass of wine at the restaurant? Your hangover the next day will be a lot more unpleasant than it was in your twenties or your thirties, partly because your liver can no longer process alcohol as effectively as it used to.

The time of life when we are obviously aging and aged—not the subtle shifts that begin to age us in our twenties, but the cold, hard realities of aging that you wake up to in your fifties—is known as senescence, or "to grow old." The word "senescence" comes from the Latin spirit of old age, Senectus, who also loaned us the root for the words "senescent," "senile," and "senator." During this stage, more and more of your cells begin to show the impact of your decades of living—these are called senescent cells. During senescence, the effects of aging can be felt more readily on the surface; as time goes on, and more and more cells become senescent, you get older and older (we'll discuss this in more detail in Chapter 6).

In your fifties, you will have less muscle, bone, and fat beneath your skin as well as less collagen and elastin, the proteins that give skin plumpness and elasticity. This means that those faint laugh lines of your forties start to get a little deeper, and the skin all over your body starts to sag a bit. At some point, your period has probably become irregular, and you may begin to experience hot flashes, cold sweats, or the brain fog that accompanies the menopause transition.

In your sixties, drier skin means those "teenage" acne flares that lasted into your forties are a thing of the past. But your skin is also more fragile and age spots more common. If you're inactive, you may have achy joints because cartilage wears down, joints become less lubricated, and muscles get weaker. Your rate of digestion slows and you may experience some new GI symptoms like acid reflux or constipation. Your metabolism will also slow down a bit, which can lead to weight gain. The bladder muscles weaken and the bladder can hold less urine. Beware of a little bit of peeing when you laugh or cough. (You may also need to take more frequent bathroom breaks.) After the age of sixty-five, you are more likely to be diagnosed with Alzheimer's disease (though symptoms can begin to appear twenty years prior).

One study found that people who have a positive outlook about aging live approximately 7.5 years longer than their glass-is-half-empty peers. Fearing aging, stressing over the symptoms of aging, and worrying about the downsides of age can actually make the aging process more challenging.

In your seventies, you have fewer taste buds, and the ones that you do have are less sensitive, so you may experience changes in your appetite. You might also experience changes in your sense of smell, since the nerves in your nose are less able to pick up subtle scents. You'll notice more skin along your neck and you may discover a bit of a jowl. Your metabolism continues to slow down, while your risk of developing heart disease rises; right now it is the leading cause of death for people aged seventy-five to eighty-four.

In your eighties, your mobility may be affected; 33 percent of those over the age of eighty have difficulty walking, and more than 25 percent have a tough time getting up from a chair. You also become more likely to develop a number of chronic conditions at once (like arthritis, high blood pressure, and diabetes) and are more likely to take three or more medications.

The good news for all of us as we look into the future is that, according to research, feelings of happiness and satisfaction actually increase with age. Along with the sags and bags, aging can bring positive feelings like increased enjoyment and satisfaction. None of us has a crystal ball to tell us exactly what our futures will hold, but myriad studies of the aged show that they experience more happiness than the young.

In fact, studies around the world have consistently found that the happiest people are between eighty-two and eighty-five years old. Scientists have actually charted the rise and fall of happiness levels over the course of a lifetime, creating a curve known as the U-bend. As you can see for yourself, we're pretty happy at eighteen, and then happiness dips in our twenties and thirties, rising again in midlife.

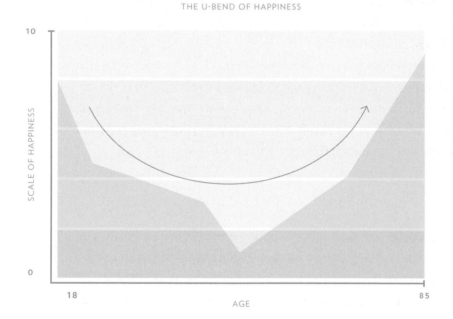

THE U-BEND OF HAPPINESS

Meanwhile, other studies show that while stress increases throughout our twenties and thirties, our self-reported stress and worry start to decrease after our forties. Our ability to manage our relationships also increases with age. So does wisdom.

The idea that there's an upside to aging is more than a spirited notion. Finding reasons to look forward to growing older is good for your physical health—and being older is apparently great for your psyche.

Today, as you know, American women can expect to live into their eighties. But a very small percentage of people will go on to celebrate their 100th birthdays. They are known as centenarians. Centenarians are fascinating subjects for scientists who study aging because they offer a model of what might be possible for all of us one day. Why do some people cruise past the life expectancy of eighty-something and just keep going? What is the magic ingredient that helps these people keep on living, and living—and living?

Some researchers have concluded that superlongevity is tied not just to physical health but also to personality. Studies of centenarians have found that people who live to be one hundred years or more are, overall, positive people. They tend not to be neurotic or overly anxious. Most exhibit conscientiousness and competency, and are willing and able to take care of life's details. They are also more likely to take care of their health. Those with a good attitude are more likely to eat well, work out, not drink alcohol excessively, not smoke, and go to the doctor. Over time, their level of function is higher than their less-conscientious peers. The centenarians studied are also generally trusting and agreeable people who are extroverted and social, always seeking connection with others. The takeaway: people who feel good about the world and about their role in it tend to live longer.

ATTITUDE ADJUSTMENT

One longitudinal study that began in 1986, the Nun Study, has found clear links between attitude and longevity. The Nun Study took a long look at the lives of a group of Catholic sisters between the ages of seventy-five and ninety-five. Researchers collected handwritten journals from the nuns—notebooks that they had kept throughout their lives—and examined them closely for emotional content. They found that the nuns who had expressed more positive emotions when they were young experienced greater longevity than the nuns who expressed more negative emotions when they were young.

Research also shows that having a good attitude about aging is tied to longevity. One study found that people who have a positive outlook about aging live approximately 7.5 years longer than their glass-is-half-empty peers.

Fearing aging, stressing over the symptoms of aging, and worrying about the downsides of age can actually make the aging process more challenging. On the other hand, cultivating a healthy view of the aging process has positive correlations for your health. People who embrace the idea of aging actually age better, both physically and emotionally.

But it's not just about being born with a lighthearted or happy-go-lucky nature. Personality and attitude are products of a combination of influences: your DNA, your environment, your social setting, and your culture. Attitudes can be communicated to us by the society in which we live and by the people around us. Even if you naturally wake up on the wrong side of the bed every morning, the positivity-longevity correlation can still be had. A positive attitude is an aspect of personality, but it's also something that you can actively cultivate by being educated, aware, and informed.

Dr. Becca Levy is a researcher at Yale who has been investigating the link between attitude and age since the 1990s. Her work has demonstrated that people who were exposed to positive attitudes about aging actually showed improvements in physical ability, like how quickly they could walk. The kinds of benefits found in the study—which arose from just letting people know that aging with health and strength is possible—were comparable to the kind of gains usually seen from exercise.

Think about that for a moment. Just learning that aging well can make you healthier.

CHAPTER 6

THE MIRROR AND THE MICROSCOPE

The Secrets of Cellular Aging, Revealed

W HEN MY NIECE WAS born a few years ago, her older siblings were already in the fourth grade and up. One of her big sisters, who until that week had held claim to the status of youngest sibling in the family, couldn't believe how smooth and velvety her baby sister felt.

"Her skin is so soft," she kept saying. "I wish *my* skin was that soft."

I said, "You're eight years old. Your skin is that soft."

She said, "No, but hers is *sooo* soft. It's so beautiful."

I did what any good auntie would and explained the circle of life.

"Look," I said. "She's beautiful. You're beautiful. Your skin used to be that soft. Auntie's skin used to be that soft. Grandma's skin used to be that soft. My skin used to be as soft as your skin, right now, at eight years old. And one day you'll have skin like mine. We're all moving forward. I'm moving forward, you're moving forward. We're like fruit on a tree. A little seed comes out, then a flower, then a fruit, and when it gets too ripe it drops off and falls on the ground. It gets eaten by animals. It becomes part of the earth."

She stared at me wide-eyed, not sure how I had made the jump from soft skin to the day we would all be devoured by wild animals.

So I bottom-lined it.

"As soon as you're born, you start to die."

She said, "Oh, Auntie!"

All the kids chimed in. "Way to go, Auntie. Bummer."

Sure, it's a bummer. We are born soft and smooth and then we age, and our skin—and everything it holds—ages along with us. Wrinkles are an external sign that within your cells, changes are taking place. The effects of age on our skin can be one of the first reminders that time is marching forward and we are going to have to hustle to keep up with it.

The quality of all your organs—the one you see and all the ones you don't—are the product of genes, choices, and environment. Just as wrinkles are the result of changes happening deep within your skin cells, heart attacks are the result of changes happening deep within your heart cells, blood cells, lung cells, and other cells.

ALL AGING IS CELLULAR AGING

What you see on the outside of your body is a reflection of what's going on inside it. Think about it—while you can't "see" the effects of aging on your inner organs, you can most definitely see the byproducts of time on your largest and only external organ: your skin. Like all your body's pieces and parts, the skin you have today is a product of both your genes and your environment. The physical surface that you see when you look in the mirror is a reflection of how your mother's skin looked at your age, along with all those long days you spent at the beach, the occasions you forgot the sunscreen, those cigarettes you had, those glasses of wine you drank, as well as how often you laughed and how often you frowned.

Wrinkles are an external result of processes that occur deep within your cells. Everyone's skin will age naturally, but some habits, like spending too much time in the sun, will speed up the effects of aging on your skin. Ultraviolet (UV) radiation dramatically accelerates wrinkles by breaking down collagen and making your skin less elastic, in a process known as "photoaging." Exposure to UV light also causes age spots—which, while harmless, can provide a very good indicator for the amount of UV damage you've accrued over a lifetime. And too much time in the sun, as we know, can also lead to skin cancer.

The quality of all your organs—the one you see and all the ones you don't—are the product of genes, choices, and environment. Just as wrinkles are the result of changes happening deep within your skin cells, heart attacks are the result of changes happening deep within your heart cells, blood cells, lung cells, and other cells. These changes are spurred by poor nutrition, lack of movement, smoking, and stress—along with genetics and the natural changes that accompany aging.

That's what we mean when we say that your health begins with your cells, because everything about you begins with your cells. The shape and structure of your heart. The pattern of your heartbeat. The quality of the blood that flows through your veins and arteries. The quality of those veins and arteries themselves. And the quality of your hair and your nails and your skin.

THE RÉSUMÉ OF A CELL

If your cells were applying for a position within your body, their résumés would have to list their specialized roles and their shared responsibilities. All your cells have specialized jobs. Heart cells beat. Muscle cells contract. But no matter the location, every single cell in your body shares the same duo of basic responsibilities that keep you alive: they make energy, they make proteins. Some of them also divide to make new cells. Throughout your years on this planet, these pathways of life are relied upon again and again and again.

With all this use and reuse, these pathways for health and sustenance also become the pathways of the aging process. But your body contains

For two weeks, a felt model of a cell sat on my kitchen table. Every morning and evening, Sandra and I would take out the pieces and move them around, examining the shape and function of the tiny organelles—the organs of your cells—that are the basis of everything we love about life. In an effort to get to know our cells more intimately, we stared at the model for days, sometimes getting paper out of the printer and finding some pencils and pens and drawing our own cells, then labeling them afterward and checking to see how we'd done. There is a little universe in every cell, a self-supporting world that takes chemicals and molecules and turns them into the stuff of life. How incredible and humbling to imagine that every breath of air, every bite of food, all contribute to the ingredients our cells need to keep us going.

One great way to get better acquainted with your cells is to grab a paper and a pen—a set of colored pencils makes it even more fun—and draw yourself a cell, organelle by organelle. Start with the CELL MEMBRANE, which is the cell's jacket. The membrane holds fluid in, keeps strangers out, and acts as a representative that lets other cells know what kind of cell they are dealing with. It is covered with receptors, special proteins that communicate with the world outside that cell and help decide who should be let in and who needs to take a seat and wait outside. Hormones, drugs, sodium, and glucose are among those seeking entry, and it is up to the membrane to decide whether to welcome them or turn them away. If anything does get in that doesn't belong, or if cellular proteins need breaking down, cells have LYOSOMES, which contain special enzymes that can digest whatever gets in their path.

Once you're past the cell membrane, you'll find yourself swimming in the thick liquid of the CYTOPLASM, and in that soft middle you will find the organelles. Just as your body has organs to process energy and get rid of waste, your cells have organelles that perform the same functions. The cell's protein-making factories are its RIBOSOMES. Proteins made by the ribosomes can be sent out of the cell to the organs and tissues that need them via the GOLGI APPARATUS, which operates as a sort of intercellular post office. If you wanted to ship something locally you'd use the ENDOPLASMIC RETICULUM, which is more like a bicycle courier.

The center of the cellular world is the NUCLEUS. The nucleus has a NUCLEOLUS, which produces ribosomes. The nucleus is also where your forty-six chromosomes live, which hold most of your DNA (your mitochondria hold a handful donated by your mother, which we'll discuss shortly).

The DNA in the nuclei of your cells hold the instructions for your genes, the ones that are expressed—which means that time or environment activates them—and the ones that are not expressed. Then there are the MITOCHONDRIA, the powerhouses of our cells. They use the oxygen in your cells to convert sugar and fats into ATP, or usable human energy. Mitochondria create the energy that is used by the cell to support all your metabolic functions.

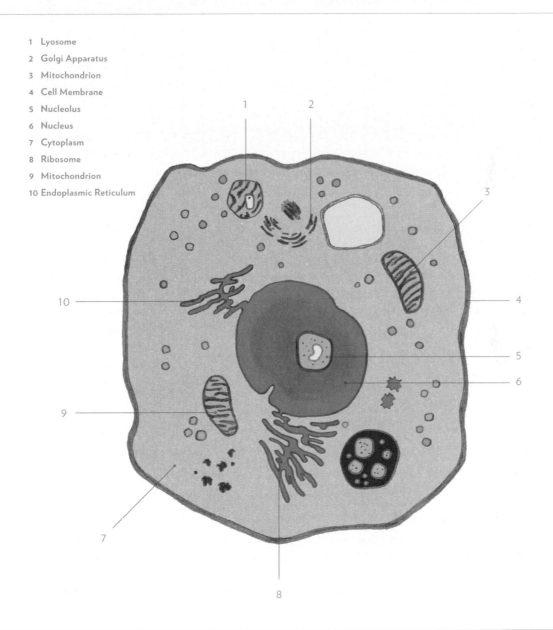

1 Lyosome
2 Golgi Apparatus
3 Mitochondrion
4 Cell Membrane
5 Nucleolus
6 Nucleus
7 Cytoplasm
8 Ribosome
9 Mitochondrion
10 Endoplasmic Reticulum

built-in protective mechanism for these pathways, tools that it can use to set your cells back on track—and minimize the damage done over time.

Most of the energy that fuels your life is produced by your mitochondria in a process known as cellular respiration. The mitochondria are components of your cells that serve as the body's energy factories. But as with most energy manufacturers, they produce a lot of waste, and if that waste is not properly disposed of, it can contaminate everything around it. The byproducts created by cellular energy production—free radicals—can damage DNA. Nearby molecules like proteins and lipids can also take a hit. All this unrest—the byproduct of our natural need for energy, remember—can lead to chronic diseases associated with aging, from heart disease and cancer to diabetes and chronic inflammation.

The fact is that as we get older, our cells become a little less able to do the essential jobs of life as well as they did when they (and we) were young. The processes by which our cells make energy, make protein, and make new cells keep us alive, but they also have the potential to create dangerous situations in the body.

Protein production is an essential cellular activity because humans are basically made of protein—we need our bodies to keep making more of it all the time so that we grow and heal and keep on living. Our DNA is a code with instructions for building many different kinds of protein, each with a specialized purpose. If you look at proteins under a microscope, you will see that they are intricately folded into unusual shapes, like little origami creatures. These specialized shapes allow proteins to do their specialized jobs correctly. As we age, new proteins aren't always folded as they should be. When the sticky insides of proteins get flipped to the outside, these misfolded proteins can bond together to form potentially harmful plaques, which have been implicated in age-related illnesses like Alzheimer's and Parkinson's diseases.

Finally, what are all the protein and energy in the world if there aren't enough cells to use them? Your body needs a constant supply of fresh, new cells to keep your organs healthy and functioning. New cells are made when an "older" cell divides to create an identical copy of itself. Each time a cell divides, it must copy its genetic material with perfect precision. Accurate cell division is essential to overall health and especially our health as we age, because cells that do not divide correctly can eventually mutate into cancer cells or other types of diseased cells.

The fact is that as we get older, our cells become a little less able to do the essential jobs of life as well as they did when they (and we) were young. The processes by which our cells make energy, make protein, and make new cells keep us alive, but they also have the potential to create dangerous situations in the body. It's incredible to imagine that every part of us is made up of these tiny little pieces, and that our health is dependent on them. But the more you understand how your cells age, the more you will understand how you can heal your body from the inside out by nurturing your cells and supporting their natural protective mechanisms.

ANTIOXIDANTS: WHY GREEN LEAFY VEGETABLES ARE THE PROTECTORS OF YOUR CELLS

If you're a consumer of healthy foods or antiaging products, chances are you've probably heard of antioxidants. They're often called out on food labeling and skin cream ingredient lists and lauded for their magical powers. Most of us have a vague idea that antioxidants are supposed to be beneficial, but have you ever wondered what they actually are, and why they're so good for us?

Antioxidants—found, for example, in vitamins E and C—help the body fight the damaging effects of oxidation, a natural process that occurs in your cells when they produce energy. While supplements containing antioxidants are often marketed as the fountain of youth, there is little scientific evidence that consuming antioxidant supplements affects chronic diseases, and some evidence even suggests that large doses of single antioxidants, including Vitamins C and E may be harmful. Instead, consistently eating vegetables, whole grains, and fruits that are naturally rich in antioxidants is your best strategy for healthy aging.

The mitochondria in your cells use 95 percent of the oxygen you breathe to power cellular respiration. But that energy-making process also places stress on our cells: this is called oxidative stress. That's because oxygen, while necessary for cellular respiration, is also dangerous to your cells. In order for your cells to survive the oxygen-rich atmosphere in your body, they must turn the waste from energy production into water. Without this neat biological trick, oxygen is toxic. You've seen the effects of oxidation on fences and old tools—rust is what happens when iron is oxidized. Want that happening to your cells? Nope, you don't.

Here's how cellular respiration works: when you eat, cells all over your body, from your muscles to your brain, devour and burn the nutrients from the food you eat that are carried to them via the blood in your circulatory system. Carbohydrates are broken down into simple sugars like glucose and eventually become ATP, which is a molecule of energy just waiting to power your cells. Your mitochondria turn that ATP into usable energy that can move your body, just as the engine of a car can turn gasoline or electricity into usable energy to move the vehicle. This usable energy is known as ADP.

When you work out, your body will respond to your increased energy needs by producing more mitochondria. The more you exercise, the more ATP you need, the more energy factories your body will build to supply that energy.

In this process of creating and using energy, your cells rely on an exchange of electrons from molecule to molecule or compound to compound (molecules made of more than one element). Although this process fuels your muscles, it also leaves some molecules or compounds with fewer electrons than they started with. A molecule that has lost an electron is unstable, hungry, and missing a piece of itself. Empty-handed, these molecules become free radicals, and will attack other molecules, like road bandits, to steal their electrons. These thefts leave other molecules without electrons, damaging them in the process. Once robbed, the victims become the aggressors, attacking other molecules until they too find the electron they are looking for.

That's where antioxidants come in. Antioxidants protect our cells and our mitochondria against the oxidative stress of electron highway robbery by doing community service: approaching free radicals and offering up their electrons. Gaining an electron stabilizes free radicals. Vitamins C and E can behave as antioxidants by protecting cells from oxidized compounds and free radicals by giving their electrons to compounds that need it. This action keeps those compounds from going out and injuring other molecules.

You've seen the effects of oxidation on fences and old tools—rust is what happens when iron is oxidized. Want that hapening to your cells? Nope, you don't.

It has been reported in the past that your skeletal muscles—which are responsible for everything from keeping your body upright to walking to nodding your head—suffer mitochondrial oxidation with age, but a new study suggests that skeletal muscle decline may be an effect of being sedentary or ill, not age. To demonstrate this, researchers compared the mitochondrial function in twelve elderly athletes and nine young athletes—and saw no evidence of major decline in skeletal muscle with age. One more reason to move it before you lose it.

The brain is at particular risk when it comes to oxidative stress and the damage that can result. While it weighs very little compared with the rest of your body, it is greedy and hungry—your brain cells require a lot of energy, all the time. Without a steady supply of glucose from your food and oxygen from your blood—your brain demands 20 percent of your oxygen-rich blood despite being only 2 percent of your body mass —your neurons die. But the free radicals that are created as a part of the energy production process can badly damage brain cells. This oxidative damage may be part of what causes Alzheimer's disease.

If you were to slice your mitochondria in half lengthwise and peek inside, you would see what could be a piece of midcentury or modern art: a beautiful maze of folds and curves. Without them we could never produce as much energy as we need to survive. But besides their aesthetic value and energy-producing skills, mitochondria are incredible for a whole host of other reasons.

First of all, *they used to be independent cells until a bigger cell swallowed them up.* How crazy is that? It is believed that back in the primordial brine, a little prokaryotic cell—a single-celled creature—got into a larger prokaryote. Instead of the smaller organism being digested, the two creatures learned to coexist. The larger organism provided the nutrition and the smaller one within it provided the energy. In this storyline, larger and larger creatures evolved from this union, and the happily ever after resulted in us.

All multicelled organisms—including animals, plants, and fungi—have mitochondria. Since mitochondria used to be independent cells, it makes sense that they have their own DNA. While there are approximately 20-25,000 total genes in the human genome (according to the Human Genome Project), only thirty-seven genes live in our mitochondria. And while the DNA contained in our cells' nuclei comes from both parents, the DNA in our mitochondria is passed down through the female lineage only. Because of this, mitochondrial DNA stays intact over generations. If you've ever tried to trace your ancestry with a cheek swab, it's your mitochondrial DNA that offers important clues.

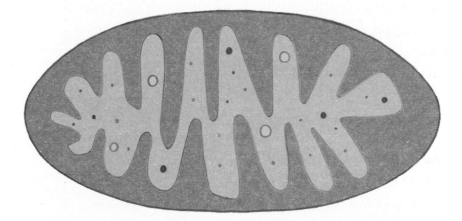

CELL DIVISION AND YOUR TELOMERES

Most cells divide. That is what they do. They do it again and again, and that is how you grow from one cell into a full adult, and that is how your tissues and organs regenerate over time, and it is how your skin heals. Cell division is everything. Each cell is a precious container of the information of your life: your DNA.

Your DNA is like a long, beaded necklace, each bead containing a tiny bit of information about you: your genes. Genes hold the instructions for making proteins and for every other cellular function, and they are organized and packaged into chromosomes. Your DNA necklace is made up of two strands of this genetic coding, all curled and coiled tightly together in the "double helix" structure. Untangled, the strand is six feet long—that's *six feet* worth of genetic information stuffed into every single one of your microscopic cells. When a cell divides, it must copy all this information perfectly. In order to do this, enzymes separate the two strands, traveling the length of each chain and creating a matching set. If there is a mismatch or a miscopy, your health could be threatened.

Every time a cell divides—which can be often, especially in places like the stomach, where our cells are replaced every five days —the ends of our DNA are especially vulnerable to wear and tear. When our DNA is copied again and again, eventually the ends can become worn and frayed—and some important genetic information can be lost. Changes in your genetic information can result in a mutation, which has the potential to develop into disease.

Luckily, our cells have special mechanisms to support our health and survival in the form of protective DNA end-caps, called telomeres. Telomeres are one of the smallest bits of our bodies but they have one of the most important jobs: to protect the fragile ends of our chromosomes, kind of like the plastic tab on the end of a shoelace that keeps the fabric from unraveling. Your telomeres take a hit every time your cells make a copy, shielding your DNA from damage that could pose a serious hazard to your health.

We have the information that we do about our telomeres in part because of the work of a scientist named Dr. Elizabeth Blackburn, who began to sequence the ends of DNA in tiny pond organisms in the 1980s. Her research, for which she and her collaborators won a Nobel Prize in 2009, showed us that just as telomeres offer protection for your chromosomes, your cells also have a built-in protection for your telomeres: telomerase, an enzyme that can help repair and lengthen our shortened and frayed telomeres.

As we age and our cells continue to divide, our telomeres become shorter and shorter. In fact, the length of your telomeres can be a measure of your health at the cellular level. Short telomeres are a sign that illness may lie ahead. When your telomeres are too short to protect your chromosomes, it's likely that your genetic material will not be copied perfectly in cell division, and cells may undergo senescence.

But it's not just cell division that shortens telomeres, a key point for those of us who want to understand how to age with better health. Scientists are now looking at the impact of environmental factors, like stress, on our cellular health as well.

To learn more about that, we visited the University of California, San Francisco, and met with Dr. Elissa Epel, a psychologist who gave us great insight into the workings of our cellular selves. Fifteen years ago, Dr. Epel

Telomeres are one of the smallest bits of our bodies but they have one of the most important jobs: to protect the fragile ends of our chromosomes, kind of like the plastic tab on the end of a shoelace that keeps the fabric from unraveling.

set up a meeting with Dr. Blackburn to discuss the topic of cellular aging from an interdisciplinary perspective: how might psychological stressors affect our cells?

Dr. Epel designed a study using two groups of moms. One group had healthy children, the other had children with a chronic illness. The researchers measured the telomeres and the telomerase of all the mothers, and discovered that women who experienced chronic stress as a result of daily caregiving had shortened telomeres and a lower amount of telomerase. In other tests, women who were the main caregivers for partners with dementia were also shown to have shortened telomeres. Additionally, women who were obese also had lower levels of telomerase. Further research has determined that telomeres can also be shortened by poor nutrition and a lack of fitness.

Women, on average, have longer telomeres than men do, and some scientists believe this is part of why women live longer than men.

Before Dr. Epel and Dr. Blackburn's collaboration, scientists widely assumed that our genes solely determine the length of our telomeres. Afterward, the world was amazed to discover that psychological stresses and lifestyle choices age us biologically, and those shifts are measurable in our cells.

What can we do to protect our telomeres? It's been shown that eating nutritious foods, getting adequate rest, and managing stress is associated with longer telomeres. Your sex also affects telomere length. Women, on average, have longer telomeres than men do, and some scientists believe this is part of why women live longer than men.

THE CELLULAR TIME-OUT: UNDERSTANDING SENESCENCE

Understanding senescence is key to understanding aging. Just as with every other part of your body, your cells become a bit less vibrant as you age. Once a cell is too old to divide properly and its telomeres have shortened past the point of effectiveness, it can become senescent. Senescence is a dormant stage that functions as a cellular "time-out." Researchers believe this pause is a crucial protector against cancer, which is the result of damaged cells dividing and creating new cells. Organisms without the protective option of senescence tend to die young, mostly because they develop cancer. At its most effective, your immune system detects and clears senescent cells regularly and your body replaces them with healthy cells capable of dividing safely and accurately.

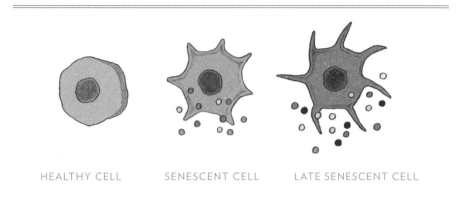

HEALTHY CELL   SENESCENT CELL   LATE SENESCENT CELL

At age fifty, however, we have many more senescent cells than we did in our youth, and they continue to accumulate as the years go by. But do senescent cells *cause* aging? That's one of the questions aging researchers are trying to answer. Some studies have shown that killing senescent cells stops mice from developing some of the disabilities associated with aging, such as cataracts and frailty.

We had the privilege of meeting with Dr. Judith Campisi at the Buck Institute. Dr Campisi is a leading expert on the links between senescence, aging, and cancer. Our meeting was so rousing that later that afternoon, her colleagues upstairs phoned her to ask what kind of party they had missed out on. Well, while they were quietly conducting research, we were having a lively conversation and peeking at a senescent cell under the microscope. We learned that while senescence protects us from mutations that may lead to cancer, senescent cells also release inflammatory substances that can ultimately speed up the aging process. Studies have shown that these substances may also contribute to the formation of tumors. At the same time, we know that the secretions of senescent cells can improve the body's ability to heal wounds and signal to the cells that make up your immune system to repair damaged tissue.

More investigation is in the works in Dr. Campisi's laboratory to help us determine all the ways that senescence affects us, and how we might best be able to reap the benefits and minimize the drawbacks of this cellular protective mechanism.

When a cell's telomeres get too short, it can become senescent, and hibernate for a while. It can also die. Some cell death is neat and orderly. Some is not. If you are injured or ill, a group of cells can implode, spewing noxious toxins into the surrounding cells and creating an inflammation zone. This process is termed "necrosis" and is responsible for irreversible death of your body's tissues. But when individual cells go awry, most cellular death occurs through the process of apoptosis, a normal biological process. "Apoptosis" is a Greek word that means "leaves falling from the trees." In this type of cell death, a cell simply follows its program to its natural conclusion—when it's time to go, it folds itself up neatly. Your immune system notices these cellular surface changes and sends white blood cells to sweep in and envelop and devour the cell—one of your body's incredible intracellular methods of reduce, reuse, recycle. With apoptosis, a dead cell gives sustenance to other cells as it goes, like a fallen leaf that nourishes the tree from which it falls.

HEALTHY CELLS MAKE A HEALTHY YOU

Each of us began our lives as a single-celled organism: a zygote, the largest of human cells. In just a matter of nine months, that cell, fueled by the energy of cellular respiration and using building blocks made of nutrition, divided millions of times and made you, a beautiful baby girl. Right now, every day, your cells are carrying out the essential functions that allow you to age for another year, and another.

The body's cellular protectors—telomeres, apoptosis, senescence—are crucial for correcting errors because, like people, cells can make mistakes. So yes, we see aging on the surface. But we never see evidence of most of the damage that can come from aging—the mistakes that can happen during cellular division and respiration—because the body is *that* good at healing itself.

Along the course of a healthy lifetime, your body is in a constant state of transformation, renewal, and rebirth at the most microscopic level. Your health begins with your cells. Every sign of aging you see and feel on the surface, every sag and wrinkle and ache and pain, begins in your cells. They are dynamic and responsive to your environment, to your stress levels, to your nutrition, and to the prompts of your hormones and the expression of your genes. Cells are born, they fulfill their destinies, and then they die, so that you can continue to live and grow older.

They are the unsung heroes of your body's biological journey through time.

CHAPTER 7

SHAPE-SHIFTER

How the Female Body Changes Through Time

MY THIGHS ARE CHANGING. One day I was drying off from my morning shower and I noticed that the skin on my thighs looked . . . different. Just like the rest of my body, my thighs are beginning to show evidence that they are merely mortal. What that means for me on a practical level is that if I don't keep working hard to maintain the muscle mass in my legs, one day I'll notice that my quadriceps feel tired and sore as I walk up the stairs with a bunch of groceries. It wasn't so long ago that I could skip a few days of working out and, while my energy level and mood might be affected, my body was still pretty resilient. It only took a few days of hard workouts to snap it back into place and regain a feeling of strength. Nowadays, it doesn't seem like anything snaps back into place just as it used to—and it certainly takes more than a few days to notice a change.

My body is different than it was even just a few short years ago. Throughout the last couple of decades I have maintained different levels of fitness. Sometimes I dedicated myself to very rigorous and time-consuming training, and other times I chose to maintain a level of fitness that was more easily integrated into my life. Each of those levels of fitness yielded different mental, physiological, and aesthetic results. I've always tried to choose my workouts based on what my body needs and when.

At this age, my body requires something more than what it used to need to feel strong. If I want to be as fit as I was ten years ago, I'd have to work much harder for it. Of course, I knew this day would come eventually, but it's still a shock that it has actually arrived. I had gotten so used to my body, and I knew exactly what I needed to do to get the results I wanted. But now, in my forties, my body is shifting in new ways. And some of those shifts are beyond my control.

The fact that all the changes of aging can seem to arrive unbidden, at once, is daunting. But the truth is, our bodies have been changing from the moment we were born. As our cells doubled and doubled again, our bodies evolved from fat, sweet babies to curious toddlers, losing the baby fat and stretching into longer-legged adolescents and then surly teens, with all the physical and hormonal shifts that accompany those years. As we became sexually mature adults, managing the responsibilities of grown-up life, the shifts in our bodies became more subtle (that is, of course, unless you've experienced pregnancy once or a few times along the way, because none of those shifts are subtle).

Our physical evolution over the course of a lifetime is supervised by our genes and our hormones, which help govern everything from our respiratory system to our circulatory system to our skin and muscles. As the years pass, the shape that our bodies will shift into depends on the health and resiliency of all our bodily systems, working collectively.

I knew this day would come eventually, but it's still a shock that it has actually arrived. I had gotten so used to my body, and I knew exactly what I needed to do to get the results I wanted. But now, in my forties, my body is shifting in new ways. And some of those shifts are beyond my control.

Thirsty, horny, hungry, sleepy—many of the physical and emotional sensations we experience every day begin as drips of hormones that send signals to our brains. In fact, our entire experience of life is influenced by an intricate dance of chemicals—including the changes our bodies experience as we age.

Hormones are chemical messengers produced by various glands and organs; their job is to communicate information to your brain and throughout the body. The word "hormone" comes from the Greek *horman*, which means to arouse or excite—and as we know, hormones can be pretty arousing and exciting. Your hormones play a critical role in many of the bodily processes

Thirsty, horny, hungry, sleepy—many of the physical and emotional sensations we experience every day begin as drips of hormones that send signals to our brains. In fact, our entire experience of life is influenced by an intricate dance of chemicals.

that keep you alive and healthy, such as building new muscle and bone, digesting food, and maintaining hydration and blood sugar levels. As we grow and develop and age, our hormone levels shift, and so do the messages they send.

Hormones are produced by your endocrine system, which is a collection of organs and glands located all over your body. Your pituitary gland—a tiny little gland in your brain that's the size of a pea—is the "master gland" of this system; it orchestrates the function of many of the other glands in the body. Behind the pituitary gland is the hypothalamus, the part of the brain that maintains your body's internal balance between all systems, known as homeostasis. The hypothalamus produces hormones and electrical impulses that send signals to the pituitary gland like a traffic cop, telling it what hormones to produce and when.

One of the hormones produced by the pituitary gland is human growth hormone (HGH). Over the course of a lifetime, the level of growth hormone in your body fluctuates; it rises during childhood, when it helps stimulate bone and cartilage growth, peaks at puberty, and then declines naturally after middle age. The pituitary gland also produces follicle-stimulating hormone (FSH), which stimulates the ovaries to release eggs for ovulation. During the menopause transition, production of FSH increases. In fact, blood tests for FSH are one method doctors use to evaluate whether you are in menopause. After menopause, FSH levels remain elevated.

Your pituitary gland is also responsible for generating thyroid-stimulating hormone (TSH). TSH regulates your thyroid, a gland in your neck that is involved in regulating everything from your metabolism to your body temperature, muscle strength, and heartbeat. When you are young, TSH is released according to the rhythm of your sleep-wake cycle, with more and more TSH being produced as the day goes on and peaking at night. As we age, that nighttime burst slows. Underproduction—hypothyroidism—can sometimes result.

Problems with the thyroid are more common in women than in men; hypothyroidism is also more common among people over the age of sixty. When the hypothalamus sends out an electrical signal instead of a hormonal signal, the pituitary gland may release antidiuretic hormone (ADH), which helps regulate water balance in the body by encouraging the kidneys to reabsorb fluids. ADH levels rise as we age, and this rise is linked to an increased risk of developing high blood pressure and heart disease. Elevated ADH levels also make elderly people more at risk for dehydration, because their kidneys become desensitized and less responsive to ADH.

Another endocrine gland in your brain, the pineal gland, secretes melatonin. You may already be familiar with this hormone, which some people take as a supplement when they have trouble falling asleep. Melatonin helps the body recognize when it is time to call it a day and get some rest. Melatonin also helps control the timing and release of female reproductive hormones, and is involved in the regulation of menstrual cycles as well as the onset of menopause.

One in eight women may experience thyroid problems during their lifetimes. As you grow older, you will probably hear more and more about thyroid disease, because as women age, our likelihood for thyroid malfunction rises. In fact, after menopause, the risk of developing clinical thyroid disease increases dramatically.

As you already know, the thyroid gland releases hormones that control many vital body functions, particularly metabolism. Thyroid issues can cause menstrual problems as well as difficulty becoming pregnant or issues during pregnancy. Women who have children are at risk for postpartum thyroiditis, an inflammation of the thyroid that affects up to 10 percent of women and can increase tiredness and irritability. It's estimated that 2 percent of older adults have hyperthyroidism (an overactive thyroid), and as many as 1 in 25 have hypothyroidism (an underactive thyroid).

SYMPTOMS OF AN OVERACTIVE THYROID INCLUDE:

• Feeling irritable, moody, nervous, hyperactive, or anxious

• Being sweaty or sensitive to high temperatures

• Having shaking hands

• Noticing hair loss

• Missing your period or getting a lighter period that usual (if you are still menstruating)

SYMPTOMS OF AN UNDERACTIVE THYROID INCLUDE:

• Issues with sleep

• Feeling tired

• Having trouble concentrating

• Having dry skin and hair

• Feeling depressed

• Being sensitive to cold temperatures

• Pain in your joints and muscles

• Frequent and heavy periods (if you are still menstruating)

With increasing age, people who have hypothyroidism in combination with high cholesterol may be at an increased risk for heart disease. And those with hyperthyroidism should be aware that they have an increased risk of developing osteoporosis.

You don't have to worry—you just have to be aware. There is a simple, routine test for thyroid diseases, as well as effective medications for these conditions. If you have hypothyroidism, thyroid hormone pills are available; as are antithyroid medications or beta blockers for hyperthyroidism. If you are experiencing any unusual thyroid symptoms, talk with your physician.

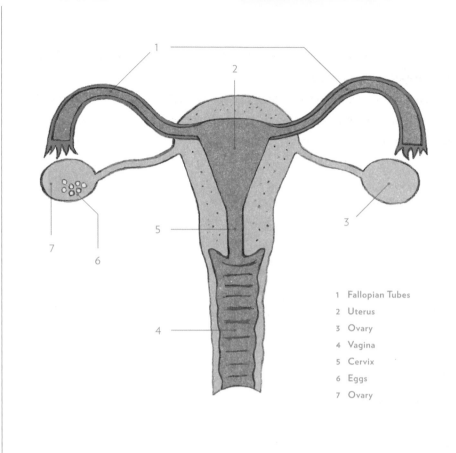

1 Fallopian Tubes
2 Uterus
3 Ovary
4 Vagina
5 Cervix
6 Eggs
7 Ovary

OUR OVARIES, OURSELVES

Your ovaries store all the eggs your body will ever produce. But your ovaries are involved in a lot more than just your reproductive ability. For one thing, they're engaged in a constant conversation with your brain. Your ovaries are busybodies—they've been chatting with your brain, about you, for your entire life.

You emerged into this world with two fully developed ovaries, one on each side of your pubic bone, like fruit attached to a fallopian tube stem, which in turn connects to your uterus. The rest of your reproductive organs are a pretty pink while your ovaries are a more minimalist gray. And these plump little hormone-makers are one of the most important organs in your body, especially as you age.

Your ovaries are a part of your endocrine system, and they produce two main hormones: estrogen and progesterone. The hormone call-and-response between brain and ovaries began when you were just a toddler. Your hypothalamus released the first drips of hormones that gave your ovaries a reason to send a hormonal message back—a conversation you experienced by becoming curious about your own body and the bodies of others. By the time you were around five years old, your hypothalamus quit sending that kind of curiosity stimulator, and your interest in the physical slowed for a few years. During that time, you were freed up to concentrate on all the richness of life as a nonreproductive being, full of playing and learning and creating (which, by the way, you will have the opportunity to enjoy again after menopause). When you were about ten, you got another fresh infusion of sex hormones, and suddenly you noticed that some of the kids in your class were *cuuuuuute*. That was just the introductory dose before the more intense influx of hormones from your ovaries that would soon change your life, forever.

Because all along, from birth until puberty, your ovaries were waiting to spark your body's transformation from a child to a woman. Before they released any eggs, your ovaries released estrogen and progesterone, readying your body to take you from the planet of childhood to the new galaxy of young womanhood, with all its accompanying blessings and challenges. As these hormones were gradually released, your hips got wider, your breasts got bigger, your pubic hair began to grow. As you became a young woman entering her reproductive phase, nature also made sure to give you full cheeks, glowing eyes, and shiny hair. Then your monthly cycle arrived.

That cycle brought with it waves of hormones—doses of estrogen, progesterone, and testosterone—that affected how you felt at different times of the month. It also helped keep you in fighting shape. Your brain was sharp. Your arteries were healthy and clear. In fact, until sometime in your thirties, you probably didn't look very different on the outside than you had just a few years before.

But make no mistake: your body was slowly shifting on the inside. Your estrogen, progesterone, and testosterone levels began to decline as your body prepared itself for the next round of shifts to come in your forties and fifties. In your thirties, you might have noticed that you had a bit less energy or, perhaps, a bit less of a drive for sex. In your forties, as your hormone levels continue to decline, so might your moods, your sleep quality, and your libido—while your risk of developing some of the diseases of aging begins to rise.

The decline of estrogen precipitates the beginning of the menopause transition, about which we will have a long and thorough heart-to-heart in Chapter 8. This is an especially important conversation for women to have with each other and with our physicians, because the menopause transition is not just about the decline of fertility. Waning estrogen levels affect our bodies in a variety of ways, including everything from storing more belly fat to changing moods and emotions to being at a greater risk for heart disease. The whole body and your changes that accompany the menopause transition will have an effect on your overall health and well-being.

AGING AND FERTILITY

Around the time you started ovulating, you had a reserve of about three hundred thousand eggs, which is a lot, but not nearly as many as the one to two million eggs your ovaries contained at birth. Of the million-plus you were born with, only a few hundred are chosen for ovulation over the course of a woman's lifetime. Every month, like clockwork, eggs are released, they wait for fertilization, and if they are not fertilized, you get your period, and the cycle begins again.

By the time a woman reaches her midthirties there is a noticeable decline in the quantity and quality of her eggs. But just as no two women age in exactly the same way, no two women experience reproductive shifts in exactly the same way. Some women will have a difficult time with fertility in their early thirties, while others may be able to conceive into their forties. As with your overall health, your reproductive health is a reflection of both your genes and your environment—there are many variables that can affect the quantity and quality of a woman's eggs. But for all of us, aging is accompanied by a decline in fertility, and as our fertility wanes, our hormones fluctuate accordingly.

HOW HAVING KIDS (OR NOT)
AFFECTS YOUR HEALTH

Like all of our life choices (and those we don't exactly choose), having children affects our health profiles—and so does not having children.

If you've experienced pregnancy, you know firsthand that those nine months change your body in myriad ways—some less pleasant than others. But research shows that having one or more children also changes your body in some wholly positive ways—including reducing your risk for breast and ovarian cancer. A study published in the highly respected journal, *Lancet*, reanalyzed 47 epidemiological studies conducted in 30 countries, totaling more than 50,000 women with breast cancer and around 100,000 control subjects. Researchers found that each childbirth reduced a woman's risk of developing breast cancer by 7 percent and every year of breastfeeding reduced the risk of breast cancer by 4 percent. Similarly, the Nurses' Health Study found a 2 percent decreased risk for ovarian cancer for every month of breastfeeding. This may be because lactation stops ovulation, reducing levels of estrogen and progesterone which are major risk factors for breast cancer. Lactation also changes how breast cells behave, making them less likely to develop into a breast cancer cell.

Women who have never given birth to children have different health risks from women who have given birth. Women who do not conceive get their periods consistently throughout their lifetimes, which can lead to a higher risk of endometriosis, a condition in which the endometrial tissue that normally lines the uterus spreads into other organs and can cause heavy bleeding, pain during intercourse, intense cramping, and in some cases infertility. Women who have had children have lower rates of endometriosis, as well as lower rates of polycystic ovary syndrome and uterine fibroids, but they do have a risk of uterine prolapse (where the pelvic floor muscles are so weakened that the uterus slips into the vaginal canal) and other birth-related conditions.

Being a mother does come with some health risks, particularly in relation to emotional well-being. It is thought that up to 1 in 7 new mothers suffer from postpartum depression. A recent study out of Germany showed that the first year after childbirth was worse for a person's psychological wellbeing than divorce, unemployment, and even death of a partner. You already know that depression and emotional stress can increase many other health risks, including threats to your longevity, so it's important for all new mothers to take care of their psychological health as well as physical health.

The most common cancer among women today is breast cancer. One in eight women will be diagnosed with breast cancer in their lifetime, equating to about 230,000 women each year in the US.

HOW THE FEMALE BODY AGES ACROSS SYSTEMS

While it sometimes seems as though all the changes of aging show up overnight—on our faces, on our skin, on our thighs—the truth is, of course, that none of the changes we experience over the years are sudden. Shifts begin slowly and subtly in our cells and in our organs; as we get older, the effects of wear and tear on our individual bodily systems show up in much more obvious ways. Here's a brief overview of how the passing years affect every system in our bodies.

RESPIRATORY SYSTEM:
YOUR LUNGS, TRACHEA, AND AIR PASSAGEWAYS

How It Ages:

Your respiratory system supplies your blood (and thus, your cells) with oxygen, and removes carbon dioxide from the body. As you get older, your lungs lose some elasticity, and your alveoli (air-containing cells) become baggy so air can get trapped, which means that you absorb slightly less oxygen into your blood vessels from the air you breathe. With age, the lungs become less able to fight infection, and your coughs become weaker and less able to help clear the lungs.

SCREENING FOR CANCER

When cells become abnormal, and those abnormal cells grow and divide unchecked, the result can be cancer. With older age comes an accelerated risk for cancer, although men are slightly more at risk than women. But for both sexes, cancer is the second leading cause of death in the United States today.

In developed countries, up to one-third of cancers are related to being obese, being sedentary, not getting adequate nutrition, and tobacco use. Lung cancer is the leading cause of cancer death in America. Of the 160,000 people in the United States who died from lung cancer in 2014, 80 percent of the cancers were caused by smoking—that's 125,000 preventable deaths every year.

There are things you can do to help protect yourself from a cancer diagnosis: quit smoking, eat right, exercise, and get screened specifically for those cancers for which you are most at risk, at the recommended ages. Screening for cancer can help physicians detect abnormal cells early, and early detection and treatment has been found to improve your odds of beating cancer.

• Ages 21–65: Cervix—get regular Pap tests and HPV tests to help protect against cervical cancer.

• Ages 50–75: Colon—get tested for colorectal cancer with a colonoscopy.

• Ages 50–69: Breasts—get an annual mammogram to screen for breast cancer.

• Heavy smokers, ages 55–80: Lungs—get a low-dose helical computed tomography to screen for lung cancer.

How to Protect Your Lungs:

If you smoke, quitting is the best thing you can do for your lungs as you age. You can also choose to get regular cardiovascular exercise and eat plenty of foods that are high in antioxidants to protect your lung cells. Air pollution is another issue—burning wood fires indoors regularly can irritate your lungs. Paint fumes, dust, particles—all of these can exacerbate or cause lung conditions. And get your flu shot. For optimal lung health, respiratory infections are best avoided.

How It Ages:

Your eyes observe by focusing light through the lens onto the retina and transmitting information about the world to your brain. As women age, our eyes take a direct hit. Dry eye affects twice as many women as men over fifty years of age. Cataracts, which cloud the eye's lens, are more likely to occur in women than in men. As you get older, you may be less able to read or perceive objects close to you, and you may experience impaired color perception and need brighter light when reading.

How to Protect Your Eyes:

Spend time away from the bright light of your computer or TV screen and wear sunglasses that protect against UV light when you're outdoors. And note that diseases like diabetes and high blood pressure can also affect your eyes. It's important to drink plenty of water to keep eyes hydrated. Antioxidants like beta-carotene, vitamins C, and minerals like zinc are all important for eye health, so eat plenty of fruits and vegetables. And watch your sodium and sugar intake; too much of either can contribute to the development of eye disease.

URINARY SYSTEM: YOUR KIDNEYS AND URETHRA

How It Ages:

Your urinary system eliminates wastes and balances your blood levels of water, electrolytes, and acid. As you get older, fewer kidney cells means smaller kidneys. Less blood flows through them, and at about age thirty, they begin to filter blood less efficiently. The urinary tract also changes with age: A woman's urethra shortens and thins as estrogen levels decline (your urethra are the ducts that channel urine from your kidneys out of your body). The bladder holds less urine, and the bladder muscles weaken.

Your kidney health can also be affected by other diseases, like high blood pressure and diabetes.

With increased age comes an increased risk of cancer. The most common cancer among women today is breast cancer. One in eight women will be diagnosed with breast cancer in their lifetime, equating to about 230,000 women each year in the US, of which 40,000 will succumb to the disease. The good news is that deaths from breast cancer have been decreasing since the 1990s due to awareness, improved screening, and early detection, as well as more treatment options. There are around three million breast cancer survivors in the US today.

As our cells continually divide and copy their genetic information, the risk of imperfect copying increases. More cellular divisions with age provide more opportunities for errors to occur. Those errors can result in mutations that allow cancerous cells to develop and flourish.

Here's how the mutations work: within our genome, we have some helpful genes called tumor suppressors that stop us from developing cancers. We also have some genes called proto-oncogenes that can promote cancers. In noncancerous tissue, our genes produce active tumor suppressors and inactive proto-oncogenes (in other words, more of what helps you and less of what hurts you). If a tumor suppressor gene is inactivated because of mutation, or a proto-oncogene mutates in a way that activates it, that cell can go on to divide uncontrollably, and develop into a cancer.

In addition to accumulating with age, mutations also accumulate through exposure to environmental factors like smoking, UV radiation, and chemical toxins. We may also be born with an increased susceptibility for cancer, and this is certainly the case for breast cancer.

We had the privilege of seeing Dr. Mary-Claire King, lauded human geneticist, give a lecture on the subject as part of the 2014 World Science Festival in New York City. She did an incredible job explaining her work to isolate and understand genes called BRCA1 and BRCA2.

BRCA genes are tumor suppressors. Mutations of BRCA genes account for 5–10 percent of all breast cancers and 15 percent of ovarian cancers. If you inherit a mutation in one copy (from either your father or mother) then you have a high risk for developing breast and ovarian cancers. For families with this gene in their lineage, the family tree can show marked dangers for carriers, but half the women who inherit BRCA1 or BRCA2 mutations do not have a family history of breast or ovarian cancer.

The presence of BRCA1 genes is associated with a more than 50 percent risk for developing breast cancer and a 15–45 percent risk for developing ovarian cancer. Dr. King reminded the audience that every bit of data she shared came from real people, with real pain, and real prognosis.

Dr. King believes that all women should be offered testing for these genes when they turn 30. If women know they carry the mutation, they can take action.

How to Protect Your Kidneys:

Drinking plenty of water is important for kidney health. So is monitoring your medications. Taking too many vitamins, over-the-counter drugs, and prescription drugs can tax your kidneys, so always talk with your doctor about the full range of medicines and supplements you're taking. And quit smoking. Smoking can hurt your blood vessels so that less blood gets to your kidneys.

THE SILENT KILLER

Every time your heart beats, blood is pumped through your blood vessels. As we get older, the pressure of our blood as it flows through our vessels can change. When the pressure becomes higher than normal, high blood pressure (hypertension) results.

Your risk for developing hypertension increases with age—in fact, about 65 percent of Americans age sixty or older have high blood pressure. Some people feel just fine with hypertension and don't even realize they are suffering from it, which is why it's been termed the "silent killer." Hypertension is a major cause of strokes and heart attacks, as well as kidney disease, eye diseases, and other illnesses.

Your blood pressure is measured for diastolic and systolic numbers. Your diastolic number is a measurement of how long the heart rests between beats; your systolic number reflects how quickly your heart pumps blood. Healthy blood pressure is less than 120 diastolic (<120) and less than 80 systolic (<80), or 120/80.

There are many risk factors for high blood pressure. Those that you can't change include genetics, age, gender (the risk for women increases after menopause), and race (African Americans have a higher risk than the rest of the US population). But there are risk factors that are within your control. Maintaining a healthy weight, abstaining from too much salt or alcohol in your diet, quitting smoking, and getting adequate exercise and rest can help. Blood pressure may also be affected by drugs and supplements. If you have high blood pressure, be sure to tell your doctor about any over-the-counter drugs you take regularly—even vitamins.

CARDIOVASCULAR SYSTEM:
YOUR HEART AND BLOOD VESSELS

How It Ages:

Your heart and blood vessels are part of your cardiovascular system, which transports blood, oxygen, and nutrients throughout the body and carries waste safely away. With age, a woman's risk of developing heart disease increases, and today in the United States, cardiovascular diseases are the number one killer of both women and men.

HEART ATTACK: KNOW THE SYMPTOMS

A heart attack happens when the blood flow to the heart suddenly becomes blocked and it can't get enough oxygen to keep pumping blood. If the condition is not treated quickly, the heart muscle begins to die. Heart attacks most often occur as a result of coronary heart disease (the most common form of heart disease).

It is important for women to be aware that the symptoms of a heart attack for us are not always the same as those for men. Our symptoms are sometimes mistaken for other issues like indigestion, the flu, or panic attacks, so women don't always seek immediate medical attention when experiencing a heart attack. Women may also be dismissive of their own discomfort, assuming it's nothing serious.

SYMPTOMS OF HEART ATTACK FOR WOMEN INCLUDE:

• neck, jaw, shoulder, upper back, or abdominal discomfort
• shortness of breath
• right arm pain
• nausea or vomiting
• sweating
• lightheadedness or dizziness
• unusual fatigue

In addition to getting annual physicals with your doctor and exercising regularly to keep your heart healthy and strong, it's important to monitor how you're actually feeling and trust your instincts when something seems wrong. Immediate action is crucial to the survival of a heart attack; about 47 percent of sudden cardiac deaths occur outside a hospital, which suggests that many people don't know the warning signs and don't get help in time. When it doubt, get it checked out!

The cause of heart disease has not been entirely determined, but many factors are known to increase your risk, including age, family history, smoking, poor nutrition, and lack of physical activity. The main treatments for atherosclerosis—the clinical manifestation of heart disease—are lifestyle interventions aimed at reversing these risk factors. Drug therapies may also be prescribed. Statins, cholesterol-lowering drugs, have been a boon for the medical industry and have reduced the incidence of heart attacks by about 30 percent. However, as with most drugs, they have potentially serious side effects that can be worsened by interactions with other drugs, especially as the need to take multiple drugs becomes common as we age.

STRENGTHEN YOUR FOUNDATION

Osteoporosis is an age-related disease that causes bones to become brittle and frail ("Osteoporosis" is Latin for "porous bone"). When we're young our bones are very supple and are constantly engaged in a process of regrowth. In fact, our entire skeleton is replaced about every ten years, with osteoclast cells reabsorbing old bone and osteoblast cells forming new bone. However, as we age, the number of osteoclast cells increases, and bone breakdown subsequently overtakes bone buildup, which causes a gradual loss of bone mass. It's as though you were renovating your home and the contractors knock down more structural supports than they rebuild. Ultimately the whole house becomes less stable than it once was.

Osteoporosis rates increase dramatically following menopause, as estrogen deficiency promotes the production of osteoclasts and subsequently leads to more bone loss. Women have about four times greater risk for developing osteoporosis than men. Risk for osteoporosis seems to have a high genetic factor, but environment also plays a role. Bone loss is increased by inactivity, low vitamin D and calcium absorption, certain medications, smoking, and low muscle mass.

Having osteoporosis is associated with an increased risk of broken bones. What's more, accidental falls and subsequent hospitalization are a major cause of disability and mortality among the elderly.

There are no symptoms of osteoporosis. Often, women realize they have osteoporosis only upon their first fracture, but you can ask your doctor for a bone mineral density test to see if you are at risk. There are a number of treatments for osteoporosis, including bisphosphonates, which help protect bones by inhibiting osteoclasts. Hormone therapy can also reduce osteoporosis for women with low bone-mineral density.

The risk for osteoporosis can be minimized by supplementing for low vitamin D and calcium and by exercising; in particular, by engaging in weight-bearing activities that build muscle to protect your bones.

Eat well, get plenty of exercise, quit smoking—and know the signs of a heart attack for women (see page 103).

MUSCULO-SKELETAL SYSTEM: YOUR MUSCLES AND BONES

How It Ages:

When we are young, our bodies continuously build bone mass, but after the age of thirty-five, we no longer have the ability to create it. In fact, we begin to lose bone mass, making our bones less dense and more fragile. We also naturally lose muscle tissue after the age of thirty-five, leading to reduced strength, less support for our weakened bones, and reduced flexibility as our ligaments and tendons become less elastic. And with the loss of estrogen after menopause, bones become weaker and more likely to break.

How to Protect Your Muscles and Bones:

Eat foods high in calcium and vitamin D to help strengthen your bones. Weight training builds muscles, which support and protect your bones. While muscle mass decreases after thirty-five, you *can* still build new muscle if you put in the effort!

INTEGUMENTARY SYSTEM: YOUR SKIN, HAIR, AND NAILS

How It Ages:

Your integumentary system includes your skin, hair, nails, and sweat glands. Your skin and hair have more than just a cosmetic role to play in your health; this system covers and protects your entire body and helps control body temperature.

As we all know, age affects the quality and color of our hair. It can turn gray or white or silver; it can change texture, becoming coarser, brittler, or thinner. But these changes don't just apply to the hair on your head—your eyelashes can also grow thin or brittle. Some elderly women even lose pubic hair and armpit hair.

Decreased collagen production means that our skin becomes drier, less elastic, and more prone to wrinkling as we age. In addition, as the fat layer under the skin thins, our tolerance for cold decreases (remember how your grandma always wore a sweater on days that didn't seem that cold to you?) and our risk of heatstroke increases. Our skin is more fragile and prone to cuts and bruises, and it can also take up to four times longer to heal from these everyday inflictions as it did in our youth.

How to Protect Your Skin:

There's plenty you can do to protect your skin. If you smoke, stop immediately. Some moisturizing products do help hydrate the skin—as does drinking plenty of water. Always wear sunscreen as well, as your skin becomes more prone to sunburn as you age.

HAIR WHERE????

You're looking in the mirror, putting on some sunscreen, let's say, when the sun suddenly illuminates your chin . . . which has at least one—oh no, wait, three—little hairs on it. And they aren't fine hairs—they're stiff little suckers, proper hairs. Or maybe you were in the shower and the hair was on your boob. Yup, that happens.

No, it's not called "ladybeard," although that would be a totally fine name. It's called hirsutism, and it's a condition in which women develop hair (usually pigmented and stiff) in places where guys typically have hair—their face, chest, and back.

This hair growth is a result of excess male hormones, particularly testosterone. It can occur in postmenopausal women if androgen levels increase at the same time as estrogen levels decrease, meaning a top-heavy testosterone-to-estrogen level in the body. It can also occur in younger women through the inheritance of diseases or the development of traits that skew their testosterone-to-estrogen ratio, such as polycystic ovary syndrome. If you notice some spry little hairs on your chin, you can take DIY action, or turn to the professionals. Plucking is effective in the short term; for a more permanent solution, you can try laser treatments or electrolysis. Or talk to your doctor about medical treatments, including a cream that can be used topically to decrease unwanted hair growth.

DIGESTIVE SYSTEM: YOUR MOUTH, ESOPHAGUS, STOMACH, GALLBLADDER, INTESTINES, PANCREAS, AND LIVER

How It Ages:

Your digestive system performs the incredible job of ingesting, digesting, and absorbing nutrients from the food you eat, and eliminating nonabsorbed food particles from the body. While aging doesn't affect the digestive system as drastically as it does the other bodily systems, as we get older, our stomach becomes less elastic and empties food into the small intestine more slowly, so it can't manage as much food at once.

The large intestine may also slow down its job of eliminating waste, leading to fewer bowel movements and chronic constipation. The production of lactase—an enzyme present in the small intestine that helps you digest dairy—may decline too, which means you might not want to have your birthday cake à la mode when you turn fifty.

The liver loses cells with age and becomes smaller and less efficient—because of this, the effects and side effects of drugs and alcohol last longer in your system.

How to Protect Your Digestive Organs:

There are a million reasons to eat plenty of fruits and veggies and drink a lot of water, and protecting the system that works so hard to get nutrition from your food is a big one. If you don't drink enough water, get enough fluids, or eat enough fiber, you can become constipated and malnourished.

IMMUNE SYSTEM: YOUR LYMPH NODES, BONE MARROW

How It Ages:

Your immune system uses various organs throughout your body to keep you healthy, like your bone marrow, which makes blood cells, and your lymphatic system, which transports potentially dangerous substances like bacteria and damaged cells to lymph organs for disposal, employing white blood cells to escort them away, keeping you from harm. With age, our immune system begins to slow down and respond to threats less quickly than it once did. This decline in immune health is linked to a rise in cancer, pneumonia, and influenza as we age.

As you get older, your immune-system changes will affect the way your body responds to vaccines. The flu vaccine, for example, is considerably less effective for elderly people than it is for younger adults. The strength of your immune system is a signal of your overall health; in fact, scientists use the immune-system response to the flu vaccine as a data point for determining a person's biological age.

How to Protect Your Immune System:

Protect your immune system so it can protect you: eat right, exercise, get enough sleep, and reduce the stress in your life. Get your flu shot every year, especially if you are sixty-five or older; it's estimated that between 80 and 90 percent of seasonal flu-related deaths have occurred in people sixty-five years and older, and between 50 percent and 70 percent of seasonal flu-related hospitalizations have occurred among people in that age group.

NERVOUS SYSTEM: YOUR BRAIN, NERVES

How It Ages:

When it comes to the aging process, the nervous system is one of our most vulnerable systems. Your brain is the control center of your entire body. Everything about you, from your personality to your memories to your ability to breathe without thinking about it: all of that is housed in your brain. The healthier your brain is, the healthier your whole body is.

Some of the cognitive changes that occur with age can manifest as degenerative diseases, like Alzheimer's disease (which, as we'll learn, affects more women than men) and Parkinson's disease. Unfortunately, we still don't fully understand what causes these diseases.

There's a lot you can do to protect your brain, and coincidentally, these are the same activities that protect your muscles, your bones, and your moods. Being physically active, eating nutritious foods, giving yourself time for rest, creating opportunities for learning, prioritizing stress relief, and embracing social connection are all essential for brain health as you age—as well as the health of your whole body.

CONTROL + SHIFT

Nothing about you has ever been static. As an infant you couldn't do anything for yourself; then you got taller and you got stronger and you got smarter, and as a ten-year-old you could do chores and do math and run a mile without stopping. And then you grew some more, and your body changed shape, and you got even taller, and even stronger, and even smarter, until you were all grown up. In your adult years, you've gotten to know your body pretty well— you're able to reliably predict how many lunges you have to do to keep everything firm, what day of the month your boobs will start to feel sore, and when you're going to crave a doughnut.

In your forties and fifties, those patterns that you've come to rely on will shift again. The way your body looks will change. The way your body responds to your nutrition and exercise habits will change. The way you feel in your body and about your body will change. So let's get ready, because time is about to speed up, and the road ahead may have some bumps you didn't anticipate when you were blithely buying pads or tampons as if you might need a lifetime supply. Because nothing in life lasts forever—including your monthly cycle.

CHAPTER 8

THE CASE OF THE HOT FLASHES

*Investigating the Mysteries of the
Menopause Transition*

———

WHEN IT COMES TO women's health, menopause is one of the biggest mysteries there is. For decades it was referred to vaguely as "the change," because while it was obvious that women's bodies went through a profound change in later life, the medical community didn't fully comprehend the biological underpinnings of it, or how a woman's biology might affect her mental and emotional experience of the world.

Today we have more information and more clues and even more questions, as researchers are working overtime to crack the case. One of the most significant revelations we've discovered about menopause is that it actually isn't just a "change," it's a transition.

Menopause doesn't happen overnight. "The change" is not abrupt: it is a gradual transition that is accompanied by symptoms that can feel sudden. The menopause transition process can begin when you are still getting your period. One month you may skip a period, and the next month your period resumes as usual. Or you may notice a symptom along with that skipped period—but when your period comes back, the symptom disappears. The road from hot flash to skipped period or skipped period to hot flash to full-on menopause can take years. And every woman's transition is individual, with a unique timeline and unique symptoms and a whole bunch of feelings unique to that woman.

We all experience different kinds of transitions at different points in our lives, and often these transitions come with challenges like physical discomforts and emotional confusion. But ultimately, in the best of worlds, they also come with opportunities for emotional and spiritual growth. This is true of every major physical transition, from learning to walk to getting your first period to recovering from pregnancy to surviving a serious illness. And it is also true of the menopause transition.

MENOPAUSE IS PERSONAL

I'm aware that over the next decade or so, my own menopause transition will begin, and I am becoming more curious about how I will feel, both mentally and physically, when this new phase of my life begins. Being a woman is not defined by getting your period or being able to have a child. But how will I feel when the assumption I've been holding closely—the assumption that I possess the ability to create human life—is gone? How will it feel to let go? How will it affect the way I feel about my own mortality, as the transition is a clear marker of the next stage of my life?

My plan is to step forward into this new phase with an open mind and open heart, and as much courage and faith as I need to muster. I don't know what's coming for me personally, but I can look back and remember what it was like to go through any new experience in my life. Every new beginning, every letting go, has required courage and has required faith, and I know that I possess both of those qualities. And I also know that I don't have to go through it alone; I can ask for help. I can seek direction from women who have been through the transition and can offer me advice from the front lines, and I can empower myself with information.

Think about it: we've all survived every physical challenge that's been thrown at us in our lives so far because we had people who cared for us, medical professionals who guided us, and resources that informed us about not only how to cope with the changes our bodies were experiencing, but also how to heal and thrive. As science has proved, even the most challenging transition is made easier when we are armed with accurate information that allows us to navigate the unknown territory ahead with strength and grace.

Can you imagine what it would have been like to get your period without knowing in advance that someday soon you might discover blood in your underwear? You would have thought you were dying.

A few decades ago, I'm guessing that someone explained to you that you were going to start getting your period before it actually happened. Think back for a moment. During that time, a parent, sibling, friend, doctor, or teacher probably explained to you what was about to go down. Hopefully, this person did a good job letting you know that it was a perfectly natural event and that it was all going to be okay. Can you imagine what it would have been like to get your period without knowing in advance that someday soon you might discover blood in your underwear? You would have thought you were dying.

We are better able to face any new challenge when we are prepared for what is coming. So let's get down to it. What is menopause, what makes it a transition, and when and how will it happen to you?

JUST THE FACTS, MA'AM

You are officially in menopause when a full year has passed since your last period, what doctors call the final menstrual period, or the FMP. The median age of menopause is 51.4, but some women reach menopause as early as 42 and some as late as 58.1.

The menopause transition includes three stages that science calls perimenopause, menopause, and postmenopause. Perimenopause itself has two stages, early perimenopause and late perimenopause. The younger a woman is when she enters perimenopause, the longer her menopause transition may last.

The stages of menopause are not perfectly discrete. You won't get a text from your ovaries announcing that today is the day you are officially entering late perimenopause. Rather, each stage sort of gradually blends into the previous and future stage as part of the overall transition. Stages are determined by the regularity of your menstrual cycle.

The beginning of the transition, early perimenopause, is where I will be soon, so I'm starting to look at what could possibly be in my near future: skipped periods and symptoms. And I'm preparing to deal with the emotional ground that will come with those changes just like you are: by learning the facts.

THE STAGES OF THE MENOPAUSE TRANSITION

The menopause transition begins with a change to your monthly cycle—it could be a skipped period or an irregular period. If you don't track your periods carefully, or you have a history of irregular periods, these changes may be easily overlooked.

The menopause transition starts with symptoms that may be as subtle as a missed period or as dramatic as waking up for the first time in a pool of sweat in an air-conditioned room in December.

When you enter early perimenopause, your cycle has shifted enough that it is no longer predictable. Your cycles may become longer or shorter than what you're used to. For instance, if you always got your period every 28–30 days, it may start to come as early as every 25 days or as late as 32 days. Or you might skip one cycle, or even two in a row, and then your cycle will resume again as normal for a few months.

Some women will also experience symptoms that can come along with the menopause transition, like changes in mood, hot flashes, sexual issues, sweating, shivers, depressive symptoms, brain fog, and difficulty sleeping.

Late perimenopause begins when you've skipped three cycles in a row.

For some women, those three months—or even more—can go by, and then their period will resume. But once you have that three-month gap in bleeding, you're in late perimenopause. During perimenopause you may experience some symptoms like hot flashes or changes in cognitave function.

Perimenopause can last for years. The younger a woman is when she enters perimenopause, the longer it can last. Even when you have officially entered perimenopause, if you're still having a period, you may still be able to get pregnant. There have been more than a few cases of a surprise late-life baby! One physician we spoke with told us a story about a perimenopausal patient who went to an oncologist because she was worried she had a tumor growing in her abdomen. That "tumor" turned out to be an embryo.

You shift from perimenopause to menopause once you haven't had a period for twelve months. Currently there is no way to predict precisely when you might have your last menstrual period, and no way to know for certain that you are really in menopause until a year has gone by with no periods.

DEFINING THE STAGES OF MENOPAUSE

In this chapter we have outlined the stages of the menopause transition as defined by several major studies. As research into menopause receives more funding and more attention, researchers are discussing a revamped model of the traditional stages that will allow us to more closely pinpoint the physical changes that take place, from shifts in hormone levels to the rise of various symptoms. For now, however, the definitions below still provide a useful timeline for understanding the menopause transition.

- **THE MENOPAUSE TRANSITION:** The more accurate way to refer to menopause, the menopause transition is the phase in a woman's life when changes in hypothalamic-pituitary-ovarian function eventually result in the cessation of the monthly period. This transition, and accompanying symptoms like hot flashes, can last for as long as fourteen years.

- **PERIMENOPAUSE:** The phase of the menopause transition in which women notice that their periods begin to slow, quicken, or skip. Perimenopause can be accompanied by symptoms like brain fog, hot flashes, and trouble sleeping. After three months of skipped periods, a woman is considered to be in late perimenopause.

- **FINAL MENSTRUAL PERIOD (FMP):** A woman's last period, which can only be determined once a year has passed by without another period.

- **POSTMENOPAUSE:** Life after the FMP.

Your final menstrual period, or FMP, is calculated backward, after a year has gone by without a period. Why does it matter when your final period is? Because the age of menopause onset affects your health in a variety of ways that extend far beyond your fertility. It's crucial to note that postmenopause, our bodies will be at a higher risk for some illnesses. For postmenopausal women, the most common causes of poor quality of life, disability, and death are cardiovascular disease, cancer, and osteoporosis.

Early onset menopause is defined as a transition that begins before the age of forty. Early onset menopause makes women more vulnerable to many of the diseases of aging; yet at the same time, the decline in estrogen production that accompanies menopause offers some protective benefit. Lower estrogen levels are linked to a reduced risk of developing breast cancer.

If you enter your transition after the age of fifty-five, then you are experiencing later-onset menopause, which has been linked to longevity. Women who don't experience menopause until they are older have lower risk profiles for cardiovascular disease, osteoporosis, and bone fractures. But later-onset menopause also means that your ovaries have continued to produce estrogen for a few extra years, and higher levels of estrogen in the body are linked to increased risk of breast, endometrial, and ovarian cancers.

Why does menopause arrive earlier for some women and later for others? The answer, as is often the case when it comes to our health, is a combination of genes and environment. The age at which your mother's and grandmother's menopause began factors into your own, as does your ethnicity, your overall health, and your body weight. Environmental factors that affect the age of menopause onset include stress levels and smoking.

No matter when your final period happens, your lifestyle choices can help mitigate some of the symptoms and risk factors that accompany the transition.

The lack of information about menopause can be frustrating. But as women of the twenty-first century, we're actually very lucky—doctors and researchers know more about the menopause transition today than ever before. Back in the early 1990s, there was barely any research to go on. Women had been telling their doctors about their menopause symptoms for years, but doctors weren't sure how to solve their problems or even how seriously to take their experiences. No set of data existed that was complete enough to provide the information that every woman approaching her mid-forties desperately needed. Over the ensuing decades, a number of new studies emerged that have provided us with a wealth of information and insight. One of these was the Study of Women's Health across the Nation, also known as SWAN—one of the most important women's health studies ever conducted in the United States.

The objective of the SWAN study was to gather information about the biology of menopause as well as the mental, social, and emotional factors of women's experiences of midlife and menopause. An observational study, SWAN began by monitoring women between the ages of forty-two and fifty-two. Over the next decades, researchers gathered data on the health of participants over the years leading up to, during, and after menopause. By the time the women in the study were between sixty-four and seventy-four, the data was rich enough to glean some important findings. And the study is ongoing.

Here's some of what SWAN researchers, and researchers of studies that preceded SWAN have learned:

• Menopause is a transition.
• Common symptoms of the menopause transition include hot flashes, cold sweats, depressive symptoms, trouble sleeping, sexual problems, and difficulties with cognitive function.
• Symptoms can range from mild to severe and may begin independently or all at once.
• The onset of symptoms can begin before a woman experiences the first obvious irregularities in her menstrual cycle.
• Body weight, smoking, stress levels, and myriad other factors influence which symptoms a woman will experience, and how severe her symptoms will be.
• Every woman's experience of the menopause transition is unique.

THE SYMPTOMS OF THE MENOPAUSE TRANSITION

Good news! Hot flashes are real. Cognitive issues are real. The depressive symptoms that accompany the menopause transition are real. How is this possibly good news? Because during the many years when we understood very little about the symptoms of menopause, women often received little help or support from their physicians or their families when they were in the thick of it. Nobody understood that experiencing increased forgetfulness during the transition wasn't just about "getting older," or that struggles with mood are common. As with many aspects of medicine concerning female biology, the mystery of menopause remained an open case until recently.

But over the past forty or so years, researchers and doctors have learned a lot about how menopause works. That makes us lucky. Not as lucky as the generations of women who will follow us, who will possess an even greater understanding of what it means to grow older, but luckier than the generations that came before us, who experienced symptoms that made them feel awful and were sometimes met with skepticism by their healthcare providers when they asked for help.

This knowledge is a gift to all of us, because understanding the changes ahead of us can not only help us respond to them in healthy ways, it can also help us mitigate the severity of those symptoms.

So let's take a moment to look a bit more closely at some of those symptoms.

HEAT WAVE

Hot flashes and cold sweats are also known as vasomotor symptoms. Even though we experience them as temperature fluctuations, they are actually a product of how the brain regulates body temperature by causing blood vessels to either constrict or dilate. Hot flashes and cold sweats can happen at night or during the day; some women experience none or one of these symptoms, and some women hit the jackpot and have both of them. Hot flashes can last for more than seven years, especially in women who first experience them when they are in perimenopause. Body weight can also impact these symptoms: women with a higher body weight *before* their FMP experience more vasomotor symptoms than women at a lower weight. But *after* their FMP, women with a higher body weight experience fewer vasomotor symptoms.

The SWAN study also found that women's responses to these symptoms affected their experiences of them. For example, women who were extra sensitive to these symptoms, who were prone to depression, or who reported being anxious when they first noted their hot flashes actually experienced them for longer and with more intensity. Yup, the ones who were the most anxious got the most severe, longest-lasting symptoms. So even when you're sweating up a storm at an inopportune moment or tossing and turning in wet sheets, try to keep calm and carry on with the knowledge that while it's annoying and weird and uncomfortable it is also 100 percent completely normal.

We all need to be especially vigilant to take good care of our mental and emotional health during the menopause transition. Emotional health affects physical health, and women who are depressed have a greater likelihood of developing certain health issues than women who are in good mental health.

Some women experience vasomotor symptoms so intensely that that it truly impacts their quality of life and ability to function normally. For perimenopausal women experiencing debilitating hot flashes, a trip to the doctor is a must.

THE MOODY BLUES

I received an email from a friend recently in which she offered to sell me her family, husband, and dog, with two kids included, all for one low price. She wasn't really auctioning off the people she loved; she was just in the throes of perimenopause and not in the mood for any of it at the moment. She was feeling irritated and annoyed and anxious, but it was reassuring that she was at least able to be funny about it.

Depressive symptoms and menopause can go hand in hand, and can vary from mild to extreme. During the transition, women are more likely to experience psychological distress and persistent negative moods than at other times in their life. And while fluctuating hormones usually take the full brunt of the blame for negative emotions, the truth is that many other factors also play a role in our experience of menopause. Just as with every other aspect of aging, our likelihood of developing mood problems during menopause is based on the cumulative history of our experiences, our choices, and our overall health.

For some women, depressive symptoms during menopause may become a serious issue—and if depression is currently or ever has been an issue for you, please read this carefully: women who have previously experienced episodes of depression (including postpartum depression) or who have a family history of depression are at the greatest risk for becoming depressed during menopause.

We all need to be especially vigilant to take good care of our mental and emotional health during the menopause transition. Emotional health affects physical health, and women who are depressed have a greater likelihood of developing certain health issues than women who are in good mental health. Remember, aging happens across all your systems, and everything is connected. Depression raises your risk of heart disease; heart disease affects the onset of menopause; the age of onset affects your risk of various other diseases. It is crucial to pay attention to your mood and ask for help as needed to maintain your physical and your psychological well-being. Strong social connections, consistent exercise, and good nutrition can help mitigate mood issues, while stress and loneliness can exacerbate them.

BRAIN FOG ROLLING IN

Each symptom of the menopause transition is its own mini-mystery that needs to be rigorously investigated. Brain fog, forgetfulness, loss of focus—women have been talking for years about the mental sluggishness that accompanies the transition, but scientists have only recently proven what we've known instinctively all along: some women really do experience brain fog during menopause.

One study, led by Dr. Miriam Weber, a neuropsychologist at the University of Rochester, tested seventy-five women between the ages of forty and sixty to assess how well they could perform cognitive tasks like learning new information, remembering that information, and paying attention to a task for a sustained amount of time. The women answered questions about what menopause symptoms they might be experiencing, and their hormone levels were measured.

What Dr. Weber's team discovered was that menopause does have an effect on women's working memory, which affects their ability to process new information, although this effect does not appear to be solely related to hormonal fluctuations. Working memory is used all the time—when we tip at a restaurant or calculate the sale price on a dress or the rise and fall of a stock price. Attention was also an issue. Staying on top of a task like reading a long and boring book or doing their taxes was extra-hard for those in the study. Other more serious issues, like depression, anxiety, and sleep problems, were often experienced by the women who exhibited brain fog.

The SWAN study also evaluated women's cognition. Since the study monitored subjects over many years, researchers were able to determine that cognitive difficulties that were mild and transient resolved for women in early postmenopause.

GO THE !?*# TO SLEEP

Not being able to sleep sucks. Count sheep all you want, but when you're staring up at the ceiling at night, there is no quantity of fluffy imaginary friends that can make you feel better about the situation. Compound the possibility of not being able to sleep with the knowledge that getting adequate rest is increasingly important as you age . . . and you've got even more of a reason to toss and turn.

Some women going through menopause will experience insomnia, an inability to fall asleep or stay asleep at night. In one analysis, 38 percent of menopausal women had trouble sleeping. In another study, researchers investigated the relationship between sleep and menopause, particularly the effects of hormones and vasomotor symptoms like hot flashes and cold sweats. They found that the women who reported difficulty remaining asleep a few nights each week were also more likely to experience vasomotor symptoms. In addition, insomnia was linked to shifting transition stages (for example, women transitioning from late perimenopause to menopause) as well as the hormonal shifts that accompany those transitions.

Currently, women experiencing menopause-related insomnia and sleep issues are treated for sleep disturbances with traditional therapies, including over-the-counter and prescription medications. In order to encourage better sleep, there are things you can do at home, too, like avoiding caffeine late in the day, keeping your bedroom as dark as possible, creating a bedtime ritual, and avoiding smartphones or computers or televisions for at least an hour before you go to sleep. If you've already taken all these measures and you're still tossing and turning at night, talk to your doctor. Getting a good night's sleep is critical to your physical and mental health—your menopause transition is likely to be lot more pleasant when you aren't sleep deprived.

SEX AND INTIMACY

Along with the changes to your reproductive system that accompany the menopause transition, you may also experience changes in your sexual health, including how you respond to sex, how much sex you feel like having, and even how sex feels physically.

Women report varying degrees of sexual issues during the menopause transition. Among the issues women struggle with are low libido, vaginal dryness, pain during intercourse, and the inability to have an orgasm during sexual intercourse. If you're experiencing any of these symptoms, talk to your doctor. There are many over-the-counter and prescription products available to help relieve symptoms like vaginal dryness. Menopause may change the way you feel about or experience intimacy, but the end of your reproductive life is not the end of your sexual life.

In fact, research shows that years of sexual satisfaction can follow the menopause transition. A 2012 study of healthy women between the ages of forty and eighty showed that sexual satisfaction actually increased with age. The study also found that intimacy is a factor in sexual pleasure; participants who reported feeling close to their sexual partner also experienced increased lubrication, arousal, and orgasm.

So just because you might be having less fun at fifty, it doesn't mean that you won't have more fun at sixty and seventy and eighty. Keep on keeping on, ladies, because for some women, it just gets better—and better and better.

TALK WITH YOUR PARTNER

Over the course of our writing this book, nearly every conversation we had with anyone would spin its way back to the subjects of aging, the human body, and how we take care of ourselves and the people we love. One day we happened to meet a man whom we'll call Max, who told us a beautiful story about his wife's experience with menopause that really stuck with us. So we wanted to share it with you.

Max had moved to the United States from New Zealand in order to be with the woman he loved. He was very passionate about his feelings for her. He told us he's always admired his parents' long and happy marriage, and said that wanted to treat his bride with the same love and courtesy his parents had always shown each other. He was very focused on her happiness, and they were very happy together.

After he'd been with her for a while, though, he noticed that she wasn't holding his hand as often as she used to. But that was fine. Things change, right? And their relationship was as strong as ever. Then one chilly evening, he got out of bed to shut off the bedroom fan. His wife asked him to turn it back on. He asked, "But why? It's so cold, you're going to freeze."

"Just turn it on, please," she said. So he did as she requested, and went to bed with an extra blanket.

The next morning, she sat down next to him and said, "I have to tell you something."

"What is it?" he asked. He was nervous now.

"I'm going through menopause," she began to explain, and he had no idea what that really meant, so they talked at length, and she told him that sometimes she felt very warm, and sometimes she felt very cold and sweaty, taking him through her symptoms.

Max said that he was so relieved to understand what his wife was going through, because now he could help her through this transition. These days when he notices that she isn't holding his hand, or that she seems to be wiping her palms on her pants, he stops what he is doing. He reaches over. He takes her hand, takes out a handkerchief, and he wipes it for her. And in doing so, he loves her in the best way he can—because she invited him in.

Until recently, most large-scale studies of menopause have been conducted on Caucasian women. Nowadays, more researchers are exploring how the menopause transition affects women of various ethnicities in unique ways, from the age of onset to the experience of symptoms. The SWAN study found that the average age of onset was a few months later for Japanese women and a few months earlier for Hispanic women. It is challenging to wholly attribute differences like these to ethnicity, because socioeconomic factors may also have an impact. Among women of all backgrounds and ethnicities, smoking, lower education levels, and unemployment significantly correlated with a younger age of menopause onset.

Differences have also been reported in how women of varying ethnicities experience menopause. The SWAN study found that vasomotor symptoms are reported more frequently by some Hispanic women and African American women.

A recent study conducted via an online survey of more than five hundred American women attempted to validate these earlier findings by asking women of four different ethnicities—white, Asian, Hispanic, and African American—about their symptoms. The survey revealed that vasomotor symptoms were the most commonly experienced symptom in all groups except for Asian women, who tended to experience more joint and muscle pain than non-Asians. It is also intriguing to note that Asian women reported less-severe symptoms overall than the other women.

As we've discussed, previous studies have linked attitude and outlook to symptom severity and length during menopause. Indeed, in this study, Asian women reported more optimism and calmness, as compared with most non-Asian women. Clearly, more studies are needed to fully understand the impact of ethnicity on the menopause transition. But it is fascinating to note that cultural attitudes do seem to impact our experiences.

HORMONES, AGING, AND INFLAMMATION

Any conversation these days about menopause eventually circles back—or zooms straight to—the role of hormones. We talked a lot about hormones in Chapter 7, since the fluctuations in hormone levels as you age affects your overall health. In a woman's body during the menopause transition, the levels of several hormones, including estrogen and progesterone, increase and decrease irregularly. This happens as the ovaries produce less estrogen and fewer eggs.

Historically, these hormonal shifts were blamed for all the health issues women faced when it came to menopause, both the shorter-term symptoms like hot flashes and the longer-term health consequences like osteoporosis.

What we know now is that symptoms such as hot flashes may result from changing hormone levels, but the details of the relationship between hormones and symptoms remains unclear. For example, after a woman's final menstrual period, when her ovaries make much less estrogen and progesterone, some symptoms of menopause may disappear, yet others may continue or get worse.

Today, some of the most cutting-edge menopause research being conducted is examining the role of hormones. But it's not the role you might expect. In fact, the current focus for menopause research is the possibility that hormone fluctuations are just another side effect of the real instigators: inflammation and aging.

For the past twenty years, the role of hormones has been the subject of much research, much discussion, and much controversy. Hormone therapy, which used to be called hormone replacement therapy, uses estrogen and progesterone supplements to alleviate symptoms of menopause. The jury is still out on the efficacy and safety of hormone therapy overall. Studies suggest that hormones have helped reduce the severity of symptoms for some women with specific health profiles. For other women, research has linked hormone supplementation to health issues like heart disease, breast cancer, and stroke. (It's important to note that this link is related to the age at which hormone therapy begins.) Where you are in your transition and whether or not your menopause occurred naturally or as the result of medical intervention may affect the safety and effectiveness of hormone treatment, as does the mix of hormones you are prescribed.

Today, some of the most cutting-edge research being conducted on menopause is about the role of hormones. But it's not the role you might expect. In fact, the current focus for menopause research is the possibility that hormone fluctuations are just another side effect of the real instigator:

inflammation and aging (we will discuss the link between aging and inflammation in great detail in Chapter 10—stay tuned). Just as geroscience researchers are exploring aging as the underlying mechanism for age-related diseases like cancer and dementia, menopause researchers are investigating the relationship among aging, the symptoms of menopause, and the underlying mechanism of cellular inflammation, which generally increases with age. Fluctuations in estrogen affect immune cells, and may exacerbate normal aging processes, which has been correlated with the increased risk of osteoporosis and other chronic diseases after menopause.

And the plot thickens.

STRENGTH IN NUMBERS

The menopause transition affects every area of your life, including the very things that make you feel like *you*: your capacity to learn and remember, your ability to sleep through the night, your desire for and physical response to sex, and your feelings about yourself, your looks, and your value in the world.

As women who, by the time we reach menopause, will likely have spent a good portion of life nurturing and caring for other people, we also deserve something else we don't always ask for: support. Support isn't just nice to have—it is essential for our health. Women with social support are happier than those without, and they are also healthier. Supportive relationships are good for us. The Nurses Study at Harvard University showed that the more friends women had, the healthier and more content they were as they aged, and researchers have shown that women with breast cancer who have close relationships live twice as long as those who don't have that kind of support. Support comes in many forms: partners, friends, family, children, professionals—and all of it matters.

One of the most critical forms of support for any woman navigating this transition comes from her doctor. Now is the time, if you don't already have a doctor you love, to find someone whom you like and respect, and who respects you. Not all physicians are aware that hot flashes, mood and sleep changes, and transient brain fog are normal parts of the menopause transition, so if your doctor looks at you doubtfully when you talk to him or her about any of these symptoms, trust your instincts and find another physician.

It's also important to open up to your partner. Menopause isn't a secret. If it is, it shouldn't be. There's nothing shameful about going through menopause, and it helps to let your partner know what you're experiencing. It's helpful for them too: if you're going through something painful or confusing that affects your behavior, your partner will notice, and if you're acting unusually, they may think it's about him or her, when really, it's about biology.

And turn to your peers. You will find that your friends are a great resource on this journey, practically and emotionally. One of the many confusing aspects of the menopause transition is how unique its symptoms are for each one of us. But the symptoms of womanhood have always been unique for each of us, haven't they? When your best friend got her period at twelve and you had to wait until you were thirteen, you secretly worried something was wrong with you. Or when you retained water and had sore boobs every month while your sister had headaches and breakouts, you wondered which one of you was "normal." Or when your friend was miserable during her pregnancies while you were happy and glowing, you wondered what she was complaining about.

As women who, by the time we reach menopause, will likely have spent a good portion of life nurturing and caring for other people, we all deserve something else we don't always ask for: support. Support isn't just nice to have—it is essential for our health.

Even though there seems to be little rhyme or reason to how our experience of womanhood compares with that of our girlfriends, thank god we've had them to talk to for all these years. From cramps to morning sickness to skipped periods to STDs to WTF is going on with my body right now, when we talk to our friends we can always find not only empathy and a shoulder to lean on, but also company that allows us to feel less isolated in our experience. There will always be common denominators that let us know that we are normal, that we are not crazy, that we are not alone. So talk to your friends. Ask them about what they're going through. Tell them what you're going through. Share information. Although not all women experience symptoms, or the same symptoms, or the same symptoms in the same way, we can always learn from and support one another.

One of the many confusing aspects of the menopause transition is how unique its symptoms are for each one of us. But the symptoms of womanhood have always been unique for each of us, haven't they?

We all need each other to survive. When you need advice, a woman who has been there and done that, a woman you want to be when you grow up, may have beautiful insights for you. And the young women in your life may have questions for you, too. At every age, we have our weaknesses and we have our strengths. When others share their wisdom with us, when we share our strengths with them, the weaknesses dissipate, and we all get stronger.

PART 3

YOU ARE HERE

The Art and Science of Living Longer

I MAY NOT HAVE GONE WHERE I INTENDED TO GO,
BUT I THINK I HAVE ENDED UP WHERE I NEEDED TO BE.

—DOUGLAS ADAMS,

THE LONG DARK TEA-TIME OF THE SOUL

CHAPTER 9

BRICK HOUSE

Building a Stronger Body with
Food, Fitness, and Rest

━━━━━

IN OUR QUEST TO discover the secrets of healthy aging, we flew across the country and back again, meeting with some of the most knowledgeable figures in aging research today. We learned about gene manipulation in worms and in flies. We observed senescent cells on a slide and visited rooms designed to measure sleep. We read about viruses and vaccines. But the most exciting discovery we made about how women can age longer and stronger is that eating well, working out regularly, reducing stress, getting enough rest—and even enjoying a glass or two of red wine—are the cornerstones of health and vitality.

This revelation came to us on the stately campus of the National Institutes of Health during a walk with a group of scientists who have dedicated their careers to understanding the mechanisms of aging. Cozy in our winter coats, we were accompanied by Dr. Richard J. Hodes, the director of the National Institute on Aging since 1993, who had just taken the time to sit down with us and, over sandwiches, explain clearly and in laypeople's terms how human life expectancy doubled. We were also with Felipe Sierra, the director of the Division of Aging Biology, who talked with us at length about longevity and genetics (in his spare time he's an artist who creates beautiful paintings and plays the guitar). And we were with Dr. Luigi Ferrucci, the chief of the Longitudinal Studies Section at the NIA and the director of the Baltimore Longitudinal Study, who gave us insight into how aging is studied. Dr.

Ferrucci is also an accomplished chef who brought his mother's recipes with him when he moved to the United States from Italy (his friends say that his home is the best Italian restaurant in town).

We share these personal details to illustrate that the scientists studying aging today are not detached, impartial observers in white lab coats. They are people who eat takeout and paint and play music and make pasta for their friends. They are human beings who have families that are affected by aging, and they themselves are affected by time, and they themselves want to stay strong and live well. What they learn influences their lives as much as ours. When it comes to aging research, we are all subjects of this grand experiment. So it stands to reason that the insights of the people on the front lines—the practical knowledge they take home with them—should be useful for us regular people to take home too.

Nutrition and fitness and rest work together to promote cellular growth, repair, and energy production, allowing all the body's systems to function optimally. You will be better equipped to manage whatever challenges come your way, whether illness or injury, if strength is on your side at the outset.

During that walk, while Sandra strolled with Dr. Ferrucci, the director of the longest-running scientific study of human aging in the United States, she took the opportunity to ask him a personal question. With all the research he had done over the years, she wondered, what did he do to manage the aging process?

Dr. Ferrucci said that every morning, he goes for a run. Despite a busy work schedule, he tries his best to eat good food. And he avoids being angry. Basically, he exercises, eats well, and manages stress. The biggest insights from the forefront of aging research today are that the simple things we've all been hearing since we were kids—to go outside and run around, to eat our vegetables and our protein, to play well with others—are exactly what we should be doing to age well. And as our grandmas always said, so is a good night's sleep.

We heard the same message over and over. Clinical physicians and researchers agree: when it comes to combating the illnesses of aging, such as heart disease and dementia, you can take a pill, or you can take a look at your lifestyle choices. Moving your body, feeding your body, resting your body, and de-stressing your body and mind are the foundations of your health.

These elements of wellbeing are often written about as if they are separate subjects, but really they are parts of a whole. Good food without fitness or sleep won't make you healthy. Working out every day but starving your body of nutrition and rest won't make you strong. Sleeping eight hours a night but not eating well or moving enough isn't going to keep you bright-eyed and alert. But if you do all these things consistently on a daily basis, you will be amazed at how your entire being responds.

Fitness and food and sleep are integral parts of a whole. Stress is a trigger that can bring the whole house down. In this chapter, we're going to focus on the way food, movement, and sleep work together to build our bodies from our cells up. In the next chapter, we're going to take a deeper look at how stress works to weaken us, and how stress relief makes us stronger.

THE TRIFECTA OF STRENGTH

The trifecta of strength—nutrition, movement, sleep—is the most important tool we have to protect ourselves as we age. Our health is largely determined by how well we balance the energy we consume, the energy we expend, and the rest we get in between all these activities. (Energy spent equals the processes that keep us alive, like our metabolism and our beating heart, plus the energy we exert by walking down the block or reading this book.) I'm not suggesting that getting a good night's sleep will cure you of malaria, or that eating a salad can make a toothache go away, or that being an athlete will keep you from getting cancer. But I *am* suggesting that nutrition and fitness and rest work together to promote cellular growth, repair, and energy production, allowing all the body's systems to function optimally. You will be better equipped to manage whatever challenges come your way, whether illness or injury, if strength is on your side at the outset.

SYMPTOM OF AGE	GOAL	HOW FITNESS HELPS	HOW NUTRITION HELPS	HOW REST HELPS
• Muscle Loss	• Build strength and muscle mass.	• Exercise puts stress on your muscles, which makes them grow. Regular exercise is crucial for building new muscle.	• Complex carbohydrates provide the energy you need for activity. Eating protein helps muscles repair and rebuild after a workout.	• Sleep allows for tissue repair and healing, which helps your muscles recover from use.
• Brain Degeneration and Memory Loss	• A sharp mind.	• Physical activity strengthens the parts of the brain that oversee thinking and memory. It also increases our brain's built-in protectors against neurodegeneration, like BDNF.	• Plant-based nutrients in fruits and vegetables offer a variety of support for brain function; omega-3 fatty acids support a healthy brain and may protect against cognitive decline; and complex carbohydrates provide fuel for brain power.	• While you sleep, your brain washes itself of sticky plaque build-up that can lead to Alzheimer's and other diseases of dementia.
• Loss of Energy	• Energy and vitality.	• Regular exercise helps build more muscle cells; more muscle cells means more mitochondria; more mitochondria means more energy.	• Complex carbohydrates and fats fuel your energy by providing essential macronutrients to your cells. Fat is a great supplier of energy because gram for gram, it provides more than twice as much energy as carbohydrates.	• Sleep energizes you and keeps your immune system robust. Getting adequate sleep also helps you make healthy lifestyle choices.
• Depression	• Better moods.	• Exercise lowers levels of stress hormones, increases serotonin production, and releases endorphins, providing an instant mood lift.	• Eating crap makes you feel like crap. Research has tied heavy sugar consumption to an increased risk of depression, and drinking soda has been linked to shortened telomeres. Sugar also contributes to chronic inflammation.	• Sleep makes us more relaxed and less stressed; lack of sleep leads to irritability and anxiousness.

Every hive of honeybees has a queen; one bee that is special. The queen is unique from her subjects in a number of ways. She has a longer body than the other bees and a much longer life span. She is the only bee in the hive with the ability to secrete a pheromone that makes the other bees work together for a common cause. And she is the only one who can reproduce.

Here's the crazy thing: genetically, she is identical to her sisters. But when they were larvae, just a mass of cells on their way to becoming honeybees, it was the queen who got the best nutrients, and the most of them. Only one thing made her more capable than the others: nutrition. In the hive, she who gets the most royal jelly gets the crown.

Across the animal kingdom, nutrition is health, it is life, it is destiny. The food your mother ate when you were a fetus influenced your mental and physical development; the food you ate growing up provided the basis for your health as an adult, and the food you eat today and tomorrow will influence how you feel as an older woman.

A World Health Organization study of more than 150,000 individuals, nearly half of them women, concluded that for people over sixty, good nutrition can extend life. What we eat affects how long we live and how we feel every day we're alive.

Your dinner choices also affect your longevity. A World Health Organization study of more than 150,000 individuals, nearly half of them women, concluded that for people over sixty, good nutrition can extend life. What we eat affects how long we live and how we feel every day we're alive. We know it to be true intuitively on a day when we've eaten well, because our stomachs don't hurt, and we have energy to work and move and be active. We also know it to be true intuitively on a day when we need to curl up for a nap in the middle of the afternoon because all our physical energy is being utilized to digest

NEVER EAT OR DRINK
• Sugary drinks • Processed foods • Late at night

SPARINGLY EAT
• Saturated fat • Dairy • Red meat • Some poultry

SOMETIMES EAT
• Fish

OFTEN ENJOY
• A glass of wine with dinner

ALWAYS EAT
• Breakfast • Lean protein at every meal • Lots of vegetables
• Plenty of whole fruits (especially berries) • Nuts
• Legumes • Whole grains • Olive oil • Fresh water

a pile of macaroni and cheese. We all love a little indulgence. But when we indulge too often, we suffer the consequences. When we feed our bodies the nutrition they need, we reap the benefits.

Many studies have shown that a Mediterranean diet—which focuses on eating fresh produce, whole grains, lean proteins and healthy fats—lowers your risk for heart disease, colon cancer, and stroke. One recent study even linked a Mediterranean diet to improved cognitive health. Researchers who tracked subjects following a Mediterranean diet over a four-year period found that participants of both sexes had reduced their risk of developing Alzheimer's disease by half! The evidence is clear: a Mediterranean diet is beneficial for your mind and body.

So how do you go Mediterranean? It's pretty simple, really. Buy a plane ticket. Or just eat more of the stuff that's good for you, less of the stuff that isn't.

MEDITERRANEAN DIET PYRAMID

MEATS & SWEETS

POULTRY, EGGS, CHEESE & YOGURT

FISH & SEAFOOD

FRUITS, VEGETABLES, GRAINS (MOSTLY WHOLE), OLIVE OIL, BEANS, NUTS, LEGUMES, SEEDS, HERBS & SPICES

After the age of fifty, our dietary needs shift to match the needs of our shape-shifting bodies. Unfortunately, this time in our lives when we need to pay extra attention to our nutrition also happens to be a time when a lot of us don't pay nearly enough attention to our nutrition. In fact, according to data from the US Census Bureau, the vast majority of baby boomers—people sixty-five and older—are overweight or obese. A full 72 percent of older men and 67 percent of older women fall into one of these categories, putting them at an elevated risk for diseases like arthritis, diabetes, and heart disease. People who are obese are more likely to become disabled and need assistance with basic living tasks as they age. They are also more likely to experience symptoms like fatigue, exhaustion, headaches, dizziness, or swelling of the legs. And they are more likely to become depressed.

> According to data from the US Census Bureau, the vast majority of baby boomers—people sixty-five and older—are overweight or obese.

On the other side of the scale, malnutrition—not getting enough vitamins or minerals—is another concern as we age. While malnutrition is not an inevitable side effect of aging, many changes associated with the process of aging—such as changes in smell, taste, appetite, and digestion—can promote malnutrition. Inadequate nutrition has a number of detrimental effects, including the loss of muscle and bone mass, as well as increased risk of illness.

While eating a Mediterranean-style diet is good advice for anyone of any age, there are some special considerations to be aware of at fifty and beyond. At this age, your risk profiles increase for many age-related diseases, so it's important to choose nutrition that supports the pillars of your health, and that includes getting enough essential minerals and vitamins. Make sure to check with your physician that you have optimal levels of these in your blood before using supplements, as too much may not be a good thing. And if you can get them through your diet and lifestyle rather than a pill, even better!

The food you eat can bolster the health of your bones—which is great news for women over the age of thirty-five, who can no longer build new bone mass. Calcium and vitamin D are a classic pair that work together to protect your bones. Your body uses vitamin D to create a hormone called calcitriol that helps it absorb calcium.

You can get calcium in dairy products, green leafy vegetables like bok choy and kale, legumes like beans, fruits like oranges and dried figs, and nuts and seeds like almonds and sesame seeds. The general guidelines for calcium intake for women are as follows: thirty-one to fifty-one-year-old women: 1,000 mg a day; fifty-one to seventy-year-old women: 1,200 mg a day. When it comes to calcium, too much can be toxic, so if you are supplementing, be sure not to exceed more than 2,000 mg a day.

You can get vitamin D in your food, in supplements, and from the sun. Just fifteen minutes of sun exposure a day gives your body the materials it needs to manufacture vitamin D. Food sources of vitamin D include egg yolks, fortified milk, and fatty fish like salmon and mackerel. The general guidelines for vitamin D intake are as follows: Under the age of seventy, at least 600 IU and no more than 4,000 IU a day. Over the age of seventy, 800 IU a day.

"D" STANDS FOR DEFICIENCY

Vitamin D deficiency is one of the most common vitamin deficiencies in populations around the world: an estimated one billon people worldwide are thought to have vitamin D deficiency or insufficiency. And research suggests that Americans are not immune to this trend: most of us have insufficient levels of vitamin D. This deficiency can have a number of consequences for your health, which range from making you more vulnerable to catching a cold to increasing your risk for developing heart disease, diabetes, and cancer. Lack of vitamin D is also thought to play a role in mood disorders like seasonal affective disorder and depression.

The easiest way to protect yourself against this deficiency is also the most fun: get fifteen minutes of unprotected sun exposure every day. Although wearing sunscreen most of the time is a good idea, SPF will keep your skin from absorbing the vitamin directly from the sun.

EAT FOR CARDIOVASCULAR HEALTH

As our nutritional needs shift to compensate for the physical changes of age, we can protect our hearts by reducing the amount of saturated and trans fats we consume, and by keeping an eye on our sodium intake.

The American Heart Association recommends that less than 7 percent of our total daily calories come from saturated fats like the ones found in red meat; and less than 1 percent of our total daily calories from trans fats, which are found in some packaged foods, fast foods, and margarine.

Sodium, found in delicious salt, is necessary for life (and taste buds) but can contribute to heart disease and high blood pressure, as well as increase one's risk for stroke and kidney disease. Processed foods and restaurant foods are two major sources of sodium. At home, go light on the salt when cooking, and be aware of how much you're sprinkling on at the table. If you are over age fifty-one, the recommended limits are about one-half of a teaspoon, or 3,200 mg of salt (equivalent to 1,300 mg sodium) a day. When you're reading the sodium content on packaged foods, be sure to count milligrams, not percentages. The recommended daily allowance on food labels is based on recommended levels for people under fifty.

EAT TO BOLSTER YOUR MOOD

Nutritional deficiencies can affect your mood. Vitamin B6, also known as pyridoxine, helps your brain make serotonin, an important neurotransmitter for positive moods. In adults, being mildly deficient in vitamin B6 is common, especially among people having trouble absorbing nutrients, who have an overactive thyroid, or who have had heart failure. Drinking a lot of alcohol also puts you at risk for B6 deficiency. Not getting enough B6 can harm your nerves, skin, and circulatory system. B6 helps form myelin, a layer of protein that protects your nerve cells from damage. Women should aim to get 1.5 mg of B6 a day: you can find it easily in carbohydrates like fortified cereals and legumes; vegetables like carrots, spinach, and peas; dairy products like milk, cheese, and eggs; and animal protein like fish, liver, and meat.

EAT TO PROTECT YOUR BRAIN

Your liver stores vitamin B12, stockpiling a reserve to supply your brain with the vitamin it needs to function properly. But after the age of fifty, your body has a harder time absorbing B12 from the food you eat, and you can become prone to B12 deficiency. Low levels of B12 are linked to Alzheimer's disease and brain-related issues from shaky movements to dementia to vision problems. B12 is mainly found in meat, dairy, fish, shellfish, poultry, and fortified cereals. Vegetarians need to be especially mindful to get enough B12. Symptoms of B12 deficiency include anemia, fever, and sweating. Women over the age of fourteen should get 2.4 mcg of B12 a day, unless you are pregnant (2.6 mcg) or lactating (2.8 mcg).

EAT TO FIGHT THE DISEASES OF AGING

The biggest health risks as we age include cancer, heart attacks, stroke, and depression. Folate, a B vitamin, offers protection against all these common diseases of aging. It can also lower levels of homocysteine, an amino acid linked to heart disease. Folate also helps protect the brain and the nervous system, and is beneficial for memory, hearing, sleeping problems, nerve and muscle pain, and depression. Folate can be found in fortified cereals, legumes, dark green leafy vegetables, and oranges. Women over the age of nineteen should get 400 mcg Dietary Folate Equivalents (DFE), unless you are pregnant (600 mcg DFE) or lactating (500 mcg DFE).

EAT TO PREVENT OBESITY

More than 60 percent of American women are overweight or obese. Obesity is linked to early death; one study found that women who were obese as they neared retirement age had a higher risk of dying early and were three to six times more likely to suffer from a disability than their leaner peers. Obesity is an underlying factor for heart disease, stroke, type 2 diabetes, high blood pressure, arthritis, and breast cancer, to name just a few diseases. Eating to maintain a healthy body weight means eating a balanced diet of fresh fruits and vegetables, protein, healthy fats, and whole grains, and cutting down on sugar and processed foods. It also means being mindful of portion sizes and keeping your energy expenditure in line with your energy intake.

ALCOHOL AND THE FEMALE BODY

A glass of wine with dinner, or even two, has been linked to a healthy heart (and a fun evening out). But keep in mind that women are especially vulnerable to some of alcohol's harmful effects. Research shows that female fruit flies experience the intoxicating effects of alcohol faster than male fruit flies do, and the same goes for us human ladies. You probably know that if you drink as much as a man does, you'll get more drunk. You may attribute this to the fact that you are a smaller weight and size. Nope. If you are drinking shot for shot with a gentleman who is exactly your weight and size, you will still feel the effects of each round faster than he will. After drinking the same amount of booze, women have higher blood levels of ethanol than men because of a gastrointestinal enzyme that breaks up ethanol faster in men.

Gastric emptying time also factors into your buzz. When a person ingests a liquid, it makes its way to the stomach, hangs out for a bit, and then it moves into the intestines. The amount of time it takes to make this trip is the gastric emptying time, and yours is slower than a man's. That half a bottle of wine sits in your stomach longer than it does for your date, which allows more time for the alcohol to be absorbed into your bloodstream.

All of this means that women should be extra vigilant about their alcohol intake and make sure to drink plenty of water when they're drinking adult beverages. Our body's response to alcohol has long-term health implications: women are more susceptible to alcoholic liver disease leading to cirrhosis. Also, a hangover sucks.

MOVEMENT BUILDS MUSCLE

Millions of years ago, when our ancestors were hunter-gatherers who relied on the natural environment to provide their caloric needs, if you didn't have the strength to go long distances on foot, to kill animals much larger and potentially fiercer than you, you were basically screwed. And even if you did have the strength to catch up to a fleet-footed animal, or the smarts to outwit one, you would still probably sustain some injuries to your muscles. By the laws of natural selection, only those who could heal from and adapt to that kind of wear and tear would survive to reproduce, and their children would also be born with those adaptable muscles.

What does that bit of evolutionary history mean for us as aging women today? It means that we've won the muscle lottery, because adaptable muscles like ours are built to withstand the forces and pressures of life. In fact, without those forces, muscles stagnate and become weaker. When you subject your body to physical challenges, your muscles develop microscopic rips and tears that are the basis of their growth. When you lift a weight, the force against your muscle is what makes it grow. Without force, without action, there is no growth.

As we age, our bodies are less primed to build muscle than they were in our youth. When you were young, your body was made to build muscle: every couple of weeks, half of the proteins that made up your muscles were automatically renewed. Then comes your thirty-fifth birthday, and renewal, while still possible, gets a little more challenging, because your tissues become less sensitive to the exertion that builds muscles. At the same time, your body's protein-producing factory starts to slow down.

Despite this decline, muscles can still get stronger as we age. Researchers have found that the process of muscle loss is dynamic and can be affected by our lifestyle choices. Add force, add action, and you've got growth—at every age.

When we visited the Buck Institute, we got to speak with Dr. Simon Melov, an Australian fitness and aging expert, and an athlete himself—we know because he showed us footage of himself doing flips in a gym (very impressive, Dr. Melov!). What was even more impressive was the astonishing MRIs he showed us of elderly people's muscles. One of those MRIs showed the muscles of an eighty-year-old man who did not work out. They were full of fat and looked like they had atrophied. Then he showed us the MRI of an *athletic* eighty-year-old. This MRI showed muscles that were red and robust, and they looked a lot like the muscles of a younger person.

Consistent fitness and nutrition can reverse the muscle loss that accompanies age. Your nutrition plays an important role in supporting your muscles, and what your muscles are hungry for is protein. But your body can't store protein over long periods the way it can store fat (or store carbohydrates as fat). Since you can't rely on protein stores, you must make it a point to eat protein at every meal to give your muscles the building blocks they need for repair and growth.

BUILDING MUSCLES PROTECTS YOUR BONES AND YOUR BRAIN

Building muscle is the best way to offset one of the other losses that comes with age: bone loss. Strong muscles support your bones. Strong muscles also protect you from falls. As we get older, our sense of balance diminishes. Falls become more common with age and more dangerous. Research shows that when an elderly person falls and breaks a hip, he or she is more likely to die than his or her able-bodied peers over the next year. In fact, up to 1 in 3 hip fractures in the elderly result in death within six months. This is especially pertinent for women: three-quarters of the roughly 250,000 hip fractures that take place each year occur in women.

That's why no matter how old we are, we have to keep moving. We've got to be active—which means going to the gym and working up a sweat on a regular basis but also just moving more often throughout the day. If you're someone who sits at a desk all day and then drives to the gym for an hour, you are still at risk for the illnesses associated with being sedentary.

If we remain sedentary, aging weakens us. But training makes us stronger. And the results for our health are astounding. A Stanford University study examined the impact of being athletic on health as we age by monitoring a sample group of runners over fifty. The study followed the runners for twenty-one years and found that they had better cardiovascular fitness, better breathing capacity, more skeletal mass, and decreased frailty as compared to their nonrunning peers. Okay, maybe that's not surprising—of course runners are in better shape than nonathletes. But the runners also had lower levels of inflammatory markers, better response to vaccines, and improved mental functioning as compared with their sedentary peers.

That's some pretty compelling evidence to get moving, if you ask me. The majority of research on exercise and aging has come to the same conclusion: fitness offers protective benefits for our mental ability as well as our physical agility at any age.

YOUR MIDLIFE METABOLISM

As women get older, our metabolisms slow. This much is clear. But how does that change affect our weight? How can a fit woman hope to stay fit, and how can a woman who is overweight or obese hope to get fit? Understanding how your metabolism works, why it slows, and how to rev it up is a great place to start.

For instance, did you know that around 70 percent of the energy your body burns every day has nothing to do with exercise? It comes from your basic life processes—your breathing, digesting, heart beating—basically all the stuff you're barely aware of that helps you stay alive. The amount of energy you burn at rest, just being alive, is called your resting metabolic rate (RMR).

From about the age of twenty, your RMR drops about 1 or 2 percent every ten years. As we age, we also tend to lose muscle mass and gain fat. Muscle burns more energy than fat, so less lean body mass means a lower RMR. This is one of those chicken-and-egg situations: do we lose lean body mass as RMR drops? Or does the decline of RMR cause the loss of lean body mass?

The mass we're talking about isn't just muscles—it's also metabolically

The majority of research on exercise and aging has come to the same conclusion: fitness offers protective benefits for our mental ability as well as our physical agility at any age.

active tissues and organs. As we age, most of our organs get smaller and our muscles get smaller, while we gain more fat. This shift in body composition contributes to a loss of energy expenditure that makes it easier for you to gain weight.

At the same age when our RMR is slipping, many of us slow down a bit

Your circulation relies on one major pump, your heart, which sends oxygen-rich blood through your arteries to be delivered to all your tissues and organs, and then pumps that blood back through your veins, which carry oxygen-depleted blood to your lungs for refueling. This amazing system stretches through miles of veins and arteries and capillaries to keep you alive. And unless you're spending most of your day in a headstand, it's usually working against gravity. Oxygen-rich blood has to make it all the way to your fingers and down to your toes—and then it has to make its way back up to your heart. That reverse journey is powered by muscles in your legs that contract and act like a pump to push the blood upward. Then there are small valves in your veins that open and close to make sure that blood keeps going in the right direction, toward your heart.

With age, your veins become less elastic, like stretchy jeans that bag at the knees. The wear and tear on the valves that control the direction of blood flow means that with age, they sometimes let some blood back into your stretched-out veins, where it pools and collects. The result can be a web of veins that show up on the surface of legs and feet as spidery lines—spider veins—or thicker, more obvious bulges known as varicose veins. Varicose veins may trouble you cosmetically, but they are important to note because they can lead to health issues ranging from pain while standing to, at the extreme, blood clots.

Aging is a risk factor for varicose veins, and so is being a woman. Women are more likely than men to develop varicose veins, possibly because changes in hormone levels during monthly cycles, pregnancy, and menopause may influence blood flow. Other risk factors include being overweight, being sedentary, and wearing high heels or tight, uncomfortable shoes.

There is no cure for varicose veins (other than cosmetic procedures) but here are some of the best ways to prevent them:

- **EXERCISE**: Keep your blood moving and keep weight in check.
- **HOLD THE SALT**: Avoiding salty food will keep your legs from swelling.
- **WATCH YOUR WARDROBE**: High heels and tightly cinched waistlines cut off blood flow.
- **PUT YOUR FEET UP**: Getting your legs over your heart eases pressure.
- **UNCROSS YOUR LEGS**: Crossing your legs can cut off blood flow.

and are less physically active. Combine the factors of less lean body mass, a lower RMR, and a more sedentary lifestyle, and *boom!* The extra pounds will just keep collecting if you don't change your habits.

Here's the good news: moving changes everything. Studies have found that older men who are very physically active can have the same metabolic rate as younger men. Specifically, endurance exercise—sustained physical exertion over a longer period of time—has been linked to maintaining a youthful metabolic rate. So get moving, and keep moving.

MOVERS AND SHAKERS

You know that aging makes you gain weight, lose muscle mass, and lose bone mass. You know that fitness can help balance weight loss, regain muscle mass, and protect your bones. You also know that age is not an excuse to *not* work out. The research proves it: fitness affects our health as we age, and it's never too late to start working out.

Fitness makes you feel good and it makes you look good. It gives us energy for work, for play, for love, for adventure. It also helps manage stress levels. Even being stuck in traffic is less stressful when you have endorphins floating in your system from a great workout. Everything is just better when we feel strong in our bodies.

Physical activity may prove to be the difference, as we age, between dependence and independence. And isn't that what we really want for ourselves in the long run? Not just to look good in a dress today—although we all enjoy feeling confident about our bodies—but to be strong and capable for the whole journey. To become the kind of older people who have the strength and energy and brain capacity to take care of our own needs, live in our own homes, make our own decisions, drive ourselves around, to see the world and keep living fully for as long as possible.

MOVE TO STAY HEART-HEALTHY

The number one killer of American women today is heart disease, and one of the best ways to keep your heart strong is to work out. Any kind of cardiovascular exercise that elevates your heart rate is a good way to protect your heart and your blood vessels and your lungs. So dance, swim, walk long distances, anything to get that blood pumping.

MOVE TO REST EASY

In one study of adults with insomnia, participants who began a consistent exercise regime reported that they could fall asleep faster, sleep longer, and sleep better than before they became exercisers. Mind-body techniques like tai chi and yoga have also been found to help people fall asleep more quickly, and sleep more soundly. And exercising in the morning has been shown to help you sleep better at night. Adding stretching and moderate-intensity workouts to a morning routine has been shown to help post-menopausal women sleep more soundly.

MOVE TO STAY SMART

Ever go for a run to clear your head? Or come up with a genius idea in the middle of a vigorous hike? It's no coincidence: exercise helps promote new connections in your brain—in your cerebellum, your balance center, and your hippocampus, your memory center. Exercise also increases neuroprotective factors in your brain, like the protein BDNF, or "brain-derived neurotrophic factor." BDNF is a neuroplasticity molecule that helps with learning and memory, and it is found in the parts of the brain that manage food and weight.

MOVE TO STAY LEAN AND BATTLE CELLULITE

We know that working out can help us manage our body weight. But can it also help us manage our cellulite? The research points to yes. Up to 90 percent of us will get cellulite at some stage of our lives—even thin women. As we age, fat cells accumulate and push up against the skin. At the same time, the biological rods that connect your skin to the muscle beneath it can strain, which combines with the fat cell accumulation to change the surface of the skin. As we age, decreased estrogen levels lead to reductions in blood flow and decreased circulation. As a result, less oxygen and fewer nutrients are delivered to skin cells. Collagen production and elasticity also decrease with age, and as fat cells increase they become more visible as cellulite under the skin.

Hips, thighs, and buttocks are particularly vulnerable spots for cellulite, as we have more fat there in the first place. Cellulite does have a genetic component, but physical activity that increases blood flow may reduce cellulite accumulation as we age.

MOVE TO KEEP MOVING

Ten percent of the middle-aged women who participated in the SWAN study reported substantial difficulties with daily activities like climbing stairs, walking a block, bathing, or dressing. The simple pleasures of life, the basic activities of caring for oneself, had become a chore for them. Another 10 percent of participants reported a lesser degree of physical limitation with activities, but still some difficulty. That's one in five women dealing with some sort of impairment—for living life. And remember, these are women in their forties and early fifties! If you need a great reason to move, move so you can keep on moving.

If you aren't accustomed to regular movement, start with the basics, like going for a walk around your neighborhood a few times a week. Add more time as your stamina builds. You can take a class, or find a trainer. Perhaps you might engage a friend who is into fitness to give you some pointers. As you get used to being more active, your body will crave it. You will develop new muscles and better balance, and your energy levels will respond: a body in motion stays in motion. And as we get older, developing new abilities will help us maintain the agility we already have.

So go out. Get going. Then go home and get a good night's sleep.

Nowadays just about everyone is well aware of the fact that smoking is deadly. There are warnings everywhere, including a pretty bold one on cigarette packaging: SMOKING KILLS.

But while we all know that smoking can kill us, the impulse to smoke can be challenging to overcome, to say the least. I know all about that. I smoked for way too long. The truth is that the entire time I smoked, I felt sick. My body always rejected it. My skin was dry, it lacked luster, it broke out. Smoking also causes wrinkles and it makes you look older. It turns your fingers yellow and your skin gray. Nothing about smoking is attractive. It's poisonous to your body and it STINKS.

If I had a time machine and I could go back to that period in my life, I would never have lit that very first cigarette. If you've never smoked, please don't start, and if you do smoke, I beg you: please, please, please stop. No matter what it takes, you must do it. You really have no other choice.

Fifteen percent of women in the United States still smoke cigarettes. Even though that sounds like a small number, smoking is the major cause of preventable death worldwide and smokers die on average ten years before nonsmokers. Smoking also reduces your fertility. It increases your risk for cancer in your mouth, esophagus, liver, kidney, and cervix. It doubles your risk of coronary heart disease. Oh, and did I mention that it STINKS? Your breath, your hair, your skin, your clothes, your home, your car, your dog, your cat. Everything stinks when you smoke. But you may not realize it, because it also dulls your sense of smell, which, in turn, dulls your sense of taste. And what is life without being able to smell the roses or enjoy a good meal?

I smoked until I was thirty, and while I knew it was "bad for me," I didn't realize exactly what that meant. Now I know that with every cigarette, I was getting older and sicker. Now I understand that age is not just chronological. That there are accelerants, like smoking. And smoking ages us on the outside and the inside. Have you ever seen a lifelong smoker? That gray, ashen quality of their skin reflects the lack of oxygen in their body, because smoking slowly strangles your body from the inside.

Tobacco smoke is full of carcinogens, substances that can cause cancer. Carcinogens affect our DNA directly, sometimes causing our cells to divide more rapidly, which can lead to unwanted mutations in your DNA. Tobacco smoke also contains carbon monoxide—you know, that stuff that you have a monitor for in your home? You have a monitor because if you are stuck in a room with a carbon monoxide leak, you can be dead in minutes. Carbon monoxide kills by taking over your body's oxygen-delivery system. As Dr. Kayvon Modjarrad explained to us, the same actions happen in your blood every time you have a cigarette. Your blood cells contain hemoglobin, which carries the oxygen in your blood throughout your body. That's how oxygen gets around. It hitches a ride on hemoglobin, your red blood cells. But carbon monoxide is more attracted to hemoglobin than oxygen is. Carbon monoxide binds more strongly and irreversibly than oxygen, and once it binds, it prevents oxygen from binding. It's like a seat on the bus that's already taken. But once that carbon monoxide is attached to that red blood cell, it's never getting off. That red blood cell is done for, forever. No matter what you do.

So here's the good news: red blood cells only last about 120 days, or about four months. So when you stop smoking, after a few months, your hemoglobin starts to regenerate. You'll have a whole new red blood cell system in about four months.

Never stop trying to quit. It takes most people who want to quit smoking a few times before it sticks. You quit, and then you go back. You quit and go back again. That's okay. Keep trying. Talk to your doctor or a therapist. Get advice. Get support. Just give it three months. All of a sudden, you're going to have an awakening. That aha moment that comes from the feeling of giving your body all the oxygen it needs and deserves. And that feeling is just going to get better and better and better and better. Trust me.

EVEN TWENTY MINUTES MAKES A DIFFERENCE.

- Try it. Say "I quit."
- Twenty minutes later . . . your heart rate drops.
- Twelve hours later: the carbon monoxide levels in your blood normalize.
- Two days later: damaged nerve endings begin to heal, and your ability to taste and smell starts to return to normal.
- Two weeks to three months later: your lung function begins to improve.
- A year later: your risk of coronary heart disease is half that of people who smoke.
- Five years later: your risk for stroke drops and becomes the same as if you had never smoked.
- Ten years later: your risk of cancer in the lungs, mouth, and throat decreases.
- Fifteen years later: your risk of coronary heart disease drops and becomes the same as if you had never smoked.
- *Twenty years later:* for women, higher risks of death are reduced to that of women who have never smoked.

If you want to live longer, stop fast-forwarding your aging, and save your own life: quit smoking.

REST TO REPAIR, REPLENISH, AND RESTORE

Our waking hours occupy so much of our energy and attention that sleep can almost become an afterthought. It can become something we squeeze in when we're not doing the things that command the top spots on our priority list: evening activities and early morning responsibilities, late nights spent out with friends and partners or late nights spent in, soothing crying children. During busy weeks, when your schedule is full and you are stretched in every direction, our sleep is often the first thing we compromise.

Yet we spend a good portion of our lives asleep—or trying to sleep—and the quality of that time spent sleeping determines our mood and our mental sharpness not just the next morning but over the next years of our life. Because sleep is not just a byproduct of being awake. It is a wholly other way of being that heals our bodies and our minds.

When you allow your body the time it needs to recover after a long day, it can perform its essential task of refreshing your cells by clearing out the waste generated by all your metabolic functions. So how much sleep is enough? A survey of more than one million Americans proved that for optimal health, seven is the magic number. Getting less than five or more than nine hours per night has been linked to a laundry list of health issues. Want to be your brightest, shiniest self today and in the years to come? Aim for seven hours of sleep each night.

RESETTING YOUR BODY'S CLOCK

You've got a clock on the wall, a clock on your phone, a clock on your wrist—and a biological clock in your body. Time lives outside us, and it also lives within us; the daily ebb and flow of energy in our bodies is known as our circadian rhythm. Your circadian rhythm is a balance of information from your internal biological clock and the light sources in your environment—including the sun in the sky and the overhead fluorescents in your office and the bulb in the lamp next to your couch and the glare of the laptop, phone, and TV in your living room. The clock inside you and the lights outside you—these are the basis of your circadian rhythm.

The rhythms of life are familiar to us all. As diurnal animals, we rise when it is light and sleep when it is dark. Humans aren't the only diurnal animals that live by this daily rhythm. Roosters crow at daylight. Some plants even close up in the night and open in the morning. But it isn't just the light that prompts us to wake and sleep: in one experiment in the 1960s, when people were kept in the dark consistently, they woke and slept according to the basic rhythms of a twenty-five-hour cycle. Which is to say, if you hid away in a cave for months, you wouldn't sleep the whole time just because it was dark. Your inner biological clock would ensure that you still operated in a rhythm that resembled day and night.

We spend a good portion of our lives asleep—or trying to sleep—and the quality of that time spent sleeping determines our mood and our mental sharpness not just the next morning but over the next years of our life. Because sleep is not just a byproduct of being awake. It is a wholly other way of being that heals our bodies and our minds.

But as we've said, your circadian rhythm is also affected by your environment. If the biological clock is the control center for your circadian rhythm, the sun (and other light) is its call to action. As the seasons change and days grow longer or get shorter, animals use sunlight to reset their internal clocks. Many species have specific cells that respond to light. Swallows have them in their heads; horseshoe crabs have them on their tails; and humans have them on their skin, even on the backs of their knees.

When we travel through time zones, expose ourselves to light at all hours of the day, and eat whenever we feel like it, we disturb these natural rhythms, along with the hormone cycles that make us feel sleepy or alert, hungry or full. That's the external result of disturbing your internal clock.

The clock is based near the optic nerve in your brain in your hypothalamus, in an area called the suprachiasmatic nucleus (SCN). It is also found in molecules throughout your body that work in concert to manipulate your circadian rhythm. The SCN is responsible for keeping your neurons and your hormones humming in tune with that rhythm. When your eyes communicate the information via your optic nerves that it's getting dark outside, the SCN tells your brain to ramp up the production of melatonin—the hormone that makes you sleepy. And shortly thereafter, you begin to yawn.

Your internal clock (or clocks, since it's a collection of cells) uses light as a signal. The light we are exposed to gives our bodies cues to release a number of hormones, including melatonin. For many centuries, this was a simple, natural process—we rose with the sun, and rested when it was dark. Today, we are exposed to light at all hours of the day. In our homes, on the street, sometimes even in the countryside—there is ambient light all around us. With all that light streaming in, our internal clocks get confused about when we should sleep and when we should stay awake. The rhythm inherent to our life cycle has been disrupted.

When that cycle is disrupted in the short term, our health is immediately affected: we might suffer from insomnia or sleepiness, become more irritable or hungrier than usual. Researchers have shown that, likely due to changes in circadian rhythms, eating when we're supposed to be sleeping (i.e., late at night) results in weight gain even if we're not consuming more calories overall throughout the day. Eating before bedtime also means that your body has to work on digesting the food you just ate while you sleep, so your digestive system never gets to take a break.

And then there's the impact of not enough sleep on our cognitive abilities. We've all experienced it: too much travel, too many late nights in a row, and we become slower, more stupid, more prone to making mistakes. Researchers call this "social jet lag." The social jet lag that arises from occasional sleep disruptions can result in cognitive impairment, causing accidents and mistakes that could otherwise have been avoided. That's the short-term consequence of messing with natural sleep patterns. But when you have a lifestyle that interrupts natural circadian rhythms on a consistent basis, the results can become even more severe. Shift work—working an overnight job—has been linked to an increased risk of heart disease. And in 2007, the International Agency for Research on Cancer (IARC) announced that jobs that disrupt the circadian rhythm are linked to an increased likelihood of developing cancer.

RESETTING YOUR CIRCADIAN RHYTHM: LIGHT MANAGEMENT

A.M. ROUTINE
Get some sun in your eyes: when you wake up, throw open the curtains, throw on a sweatshirt, and get outside for a ten-minute walk to spark your inner clock.

P.M. ROUTINE
Let your eyes rest: reduce blue light exposure at night—that means shutting down phones, computers, tablets, and other devices at least an hour before bed.

Sleep is critical to health. And it is important to realize that we need a full night's sleep—seven hours—not just bits and pieces, because sleep is not an on-off switch or something that can be collected in fragments. A full night's sleep encompasses gradual shifts through several biological stages that give us rest, give us dreams, and give our physical and emotional bodies time to recharge and heal. When you fall sleep, your whole body slows down. After a while, your eyes dance back and forth behind your closed eyelids, and you begin to dream. Your rapid eye movement (REM) sleep (the sleep of dreams) and your non-REM sleep (the quiet sleep that takes you into deeper levels of consciousness) are interdependent. You need both if you are to wake up the next day well rested.

AGING FROM A TO ZZZZZZZZZZ

Busy happens. Travel happens. Early mornings and late nights happen. And while most of us can't sleep in or take a nap every time we feel tired, we can all try a little bit harder to make sure that rest is a priority, just like our fitness and our nutrition.

SLEEP FOR MEMORIES

Sleep helps us retain long-term memories. Sleep is also the time when our short-term memories are filed away for long-term safekeeping. When we sleep, our brains can integrate new memories, downloading what we've learned and experienced from a temporary holding place to longer-term storage. Inadequate sleep gets in the way of that process. We've known for some time that with age comes loss of brain cells, trouble sleeping, and memory issues. What we know now is that these are all related. Older adults who don't sleep well don't perform well on memory tests. And as the brain's cerebral cortex shrinks with age, sleep quality gets worse, and so does our ability to remember. The Harvard Nurses' study showed that in women over the age of seventy, too little or too much sleep was associated with an earlier decline in cognitive health.

SLEEP FOR CALM

Poor-quality sleep interrupts dreaming, and when we dream, we are able to resolve neglected emotions from the day before, turning challenges to our sense of self into stories we can understand. Without dreams and sleep, complex emotions follow us to a new day. Dreaming helps us maintain our emotional balance.

SLEEP FOR BEAUTY

Dark circles under your eyes, puffy skin, red eyes . . . staying up all night is not the way to look your freshest the next morning. We all know that. But did you know that missing sleep can exacerbate some skin disorders like eczema or rosacea? Or that your skin uses the nighttime rest to heal and repair? Cuts heal faster with plenty of rest, too, and so do bruises. To feel great and look your best, get enough rest.

BALANCE IS FOOD, MOVEMENT, AND REST

If you hold on to one piece of practical information from this book, here is what we would love for that to be: exercise and food and rest are all parts of the same subject, pieces of a whole, the foundational materials of human life that will play a large part in determining the quality and number of years that we have on this planet. Eating nutritious foods, moving often, and getting enough rest are the keys to healthy aging.

Exercise and food and rest are all parts of the same subject, pieces of a whole, the foundational materials of human life that will play a large part in determining the quality and number of years that we have on this planet.

Here's how my most perfect days go: I wake up rested from a night of sleep and dreams. The first thing I do is make my bed, smoothing the covers down with purpose, because this is one of the rituals that, for me, marks the beginning of a bright new day. After I brush my teeth, I'll drink a liter of water, and then I'll meditate for twenty minutes, because it relaxes my body and brain, putting me in a calm, energetic state. Then it's time for food, some protein and some carbs and some fat—perhaps a piece of avocado toast or a bowl of savory oatmeal. After that, I work out. This whole routine takes about an hour and fifteen minutes from start to finish, and the balance of its components—rest, nutrition, and movement—is what I have found to be the perfect formula for getting me energized, excited, and ready for the day ahead.

Throughout my day, I try to be conscious of my food intake and my movement. How I feel is a good indicator of whether or not I've given myself adequate rest, nutrition, and activity. If I'm active without getting enough food, I'll feel depleted. If I eat too much and move too little, my stomach will hurt or I will feel slow and weighed down. If I move throughout the day, and I fuel that movement with consistent healthy snacks, my energy stays up.

I eat dinner on the earlier side, because I don't like to go to sleep with a full stomach. In the evening it is time to reverse that energy flow and start to wind down. So I avoid heavy foods before bedtime. I make sure my room is nice and dark. I keep all those electronics with their blinking blue and green and red lights out of my sleeping space. I create the best setting for rest I can give myself, so that I'll have energy for the next day. And in the morning, I wake up, and the process begins again.

Of course this perfect day is imaginary, because no day is perfect. But any day that I can implement some version of this formula is a better day than the ones I can't. If I don't sleep well, if I miss breakfast or eat something that looked rich and delicious on a menu but turns out be tooooo rich and delicious, if I miss my workout because I have a stack of meetings that seem more pressing at the moment—well, I suffer for it, just as we all suffer for choosing not to take care of ourselves. And only one element of the trifecta of strength is not enough, and two out of three won't cut it, either. Letting these basic needs become imbalanced hurts our hearts, hurts our brains, and speeds up the rate of aging in our cells and organs. No day is perfect, but some days seem to inch closer than others.

Only one element of the trifecta of strength is not enough, and two out of three won't cut it, either. Letting these basic needs become imbalanced hurts our hearts, hurts our brains, and speeds up the rate of aging in our cells and organs.

Nutrition. Movement. Rest. These are the threads from which our human experiences are woven, and they are the basis of our strength as we age. The more we give ourselves these essential but often overlooked or undervalued gifts, the more we give ourselves the physical strength we want, the cellular health we need, and the vitality and energy we crave today and in the days that follow.

CHAPTER 10

CHILL OUT

Support Your Immune System by Managing Stress

I WILL NEVER FORGET THE day that I was on a film set at the Los Angeles Zoo, about to deliver a monologue, when suddenly, I couldn't remember my lines. The words were just—gone. I had learned them, I'd said them a million times, but I couldn't access them. I was in a car with the windows up and no air conditioning on, under a tent, under lights, in a parking lot. It was ninety degrees in the Valley and about 1,000 degrees in the car. I could see all the grip guys, sweating, holding heavy equipment, and looking at me like, "Get your lines right, woman, before we melt."

I was so stressed. I was so tired. I knew I needed to pull it together. I've been an actor for twenty years, and I take my job seriously. I work hard when it's time to work hard, because I know I'm not the only person there who wants to go home at the end of the day. The set can be a stressful place to work on the best of days. There is the pressure of time, of having so many bodies in a small space, of having so many personalities who all need something and want something *now*. And remember that for every actor on set, there are multiple stagehands, grips, people who have put in a long, physically demanding day, people who are also under pressure and want everything to go well and seamlessly.

And there I was, holding everything up. If I hadn't been stressed before, I was stressed then. What I needed was to take a step back, to reclaim that calm part of me that knew those lines cold.

So I said, "Give me twenty minutes." I ran back to my trailer and I meditated. And then I ran back to the car.

And I nailed it. Done, thank you very much. And we were out of there.

The reason I remembered my lines is not because meditation is magic, although it can feel pretty magical sometimes. I remembered my lines because meditation helped me disentangle myself from the stressors—the heat, the pressure, the feeling that I was wasting other people's time—and find that quiet place where I was an actor with a job to do.

The ability to untangle myself from stress also has long-term benefits for my health, which is part of why I appreciate meditation so much. Like a villain in a fairy tale, stress can wreak widespread havoc if it gets you under its spell. Stress can impair your performance at work. It can encourage the accumulation of belly fat. It can completely jack your immune system so that the natural protective mechanisms that are meant to keep you balanced and healthy are thrown into overdrive and can instead cause major illnesses.

It can also make your hair turn gray.

THE VISUAL SIGNS OF STRESS

Those first subtle glints of silver in your lush, thick hair. Is that a highlight? Is that the sun catching a strand? Nope, it's your first gray.

Those initial strands of gray hair, the kind that time—not a good stylist— bestows on us, often start to crop up in our early thirties. Some say that the color of your hair can influence when the gray begins to arrive, with redheads noting the silver at thirty, brunette ladies at thirty-two, and blondes at thirty-five. Gray hair is often the first visible sign that we are getting older. For women who color their hair regularly, it can pass unnoticed for so long that by the time the gray finally shows up, they barely notice its arrival. It can send women who are proud of their "natural color" rushing to the salon. It can make any of us pause to think about the journey we're on and how many years have passed.

It can also turn hair that was kind of ordinary into something remarkable. My mother, in her sixties, has beautiful silver hair that complements her blue eyes. And these days the younger girls go gray on purpose: I've seen plenty of young women sporting silver gray hairdos alongside the blondes and blues and pinks. Gray has become a metaphor for hiding or embracing aging, for fashion or statement, but it is purely biological at its basis. Like cellulite and wrinkles, gray hair is the result of a cellular reaction to time and to your environment.

Why do we go gray? Hair color comes from melanin, which lives in your hair follicles. As you age, the stem cells in those melanin-making follicles decrease, which means that less melanin is produced. Gray hair is often inherited, so you may go gray around the same age your parents did. But, as with all the changes that come with age, lifestyle factors also influence when you go gray. Factors like stress.

> Gray has become a metaphor for hiding or embracing aging, for fashion or statement, but it is purely biological at its basis. Like cellulite and wrinkles, gray hair is the result of a cellular reaction to time and to your environment.

Stress ages our bodies, and it can show up as streaks of gray. The cells that make melanin do not flourish in an environment poisoned by stress. Stress hormones can damage melanin cells and can potentially accelerate the appearance of gray hairs.

Combined with years of other environmental tolls, stress can not only make your hair change color, it can also compromise your immune system, your body's natural protection against illness and disease. Your immune system naturally changes with age, but a lifetime of stress, environmental exposure, and poor lifestyle choices can cause the immune system to malfunction. Physical and psychological stress increases the production of stress hormones, which can suppress the ability of certain immune cells to fight infection. That's why we're more prone to getting sick after periods of stress. Conversely, stress hormones can hyperactivate other immune cells. This pumped-up immune response generates chronic inflammation in the body. An inflamed body offers the perfect environment for the aging process to speed up—a process known as inflammaging. Chronic stress accelerates inflammaging.

The more we can reduce stress, eat well, be active, and get enough sleep, the more resilient our immune systems will be as we age.

HOW YOUR IMMUNITY DEVELOPED

From the moment you took your first breath, you became a host to all sorts of microbes. As a newly minted human you were vulnerable to disease from even the tiniest exposure to germs—and yet you were also born with a powerful ability to fight disease and illness from the inside out. This resilience is housed in your immune system, which develops over the course of your life in response to your experiences and exposures.

All creatures have some form of immune system, even flies and worms, which tells us that this system is very old, evolutionarily speaking. Humans and some other animals have an "innate immunity" as well as a more advanced immune system that includes the ability to remember past illnesses or invaders, and to learn from past responses. This is called "acquired (or adaptive) immunity," because as you grow and learn about the world, so does your immune system. The adaptive immune system is distinct from the innate immune system. Adaptive immune cells make antibodies and specifically kill virally infected cells, and even tumor cells.

Your immune system has a built-in surveillance system that is constantly scanning your body for potential problems. When it identifies something as foreign, it kicks into gear. Foreign bodies—antigens—can be anything from bacteria and viruses to chemicals or pollen. If a substance is determined to be foreign and potentially dangerous, your immune system sends soldiers called antibodies to attack and, usually, destroy the invading antigen. Your body will continue to make antibodies for the attack until the antigen is beaten back and the threat is over.

Where does immunity "live" in the body? There is no one area; in fact, immunity is facilitated by a number of different organs and systems:

- **THE ENTIRE GUT** is one of the most important parts and certainly the largest part of the immune system. It is the intestinal barrier between our bloodstream and the pathogens that hop a ride with the food we eat. The tissues that line your gut contain nearly two-thirds of your entire immune system, making them the largest planet in your immune galaxy.
- **YOUR THYMUS GLAND**, which is a part of your endocrine and lymphoid system, helps create your immune response in the first few years of life; after that it shrinks and becomes an inactive organ.
- **YOUR LYMPHATIC SYSTEM**, which consists of your lymph nodes and vessels, circulates lymph fluid to transport nutrients from your bloodstream to your tissues, and to remove waste material. The nodes filter the fluid and trap bacteria and foreign particles there, where they can then be destroyed and removed by white blood cells.
- **YOUR BONE MARROW**, which is found in the long bones of your legs and arms, inside the vertebrae of your spine, and in your pelvic bone, is also a part of your immune system. This soft tissue, called red marrow, makes red blood cells, platelets, and infection-fighting white blood cells. You also have yellow marrow, made of fat and connective tissue, which produces some white blood cells.
- **YOUR SPLEEN** filters your blood, getting rid of old blood cells and platelets, or cells that have been damaged, and destroys bacteria, particularly those that have a sugar capsule around them (like *streptococcus pneumoniae*, the main bacterial cause of pneumonia).
- **A POPULATION OF HEALTHY, HELPFUL BACTERIA KNOWN AS THE MICROBIOME** lives in your digestive tract, and your body relies on them for digestive and immune support. You have about three pounds of foreign bacteria in your body—in fact, these bacterial cells far outnumber your human cells: you are roughly 10 percent human. The balance of your particular microbiome—the type and number of bacteria that are present—depends on the food you eat and the lifestyle choices you make, including any medications that you take. The connections between your immune system and your microbiome are just beginning to be explored by scientists.

AGING AND THE IMMUNE SYSTEM

For the first six months of your life, you relied on your mother's antibodies for survival. Some were transferred to you in utero; others during vaginal birth; and even more during breastfeeding, which is connected to greater immune health and longevity. As you acquired her antibodies, her immune system's memories became your immune system's memories. As you grew into a toddler and were introduced to new foods and new environments, your immune system determined what was okay and what might be dangerous.

Your immunity continued to grow and develop over the course of your childhood, learning with every exposure, growing and becoming more resilient. Once your immune system identified a foreign substance, it could recognize it more easily and respond more quickly to it the second time around. With each exposure, your immune system gained a bit more knowledge, reinforcing tolerance or creating antibodies to fight an invader. Over the years, as you reached maturity and traveled and tried new foods and were exposed to new illnesses and diseases, you created your own memory bank of immunity.

With weakening immunity on the horizon, it is extra important to do everything we can now to make ourselves stronger. That includes developing strategies to help us manage stress before it becomes a danger to our physical and emotional resilience.

Around age fifty, the immune system—especially the adaptive immune system—begins to experience a natural weakening that can be compounded by unhealthy choices and stress, or bolstered by healthy living. Signs of a weaker immune system range from cuts that take longer to heal to being sick more often each winter. Older people are more susceptible to coming down with minor infections, like the common cold, as well as to chronic ones, such as pneumonia and the flu. Once you have a weakened immune system, your body is also less capable of mounting an effective counterattack against these invaders. All of this means that as you age you are likely to get sick more often, and for longer periods of time, as compared with when you were younger.

The consequences of decreased immunity can extend far beyond a bad case of the flu. For instance, as you get older, if your immune system doesn't function well, it is also less able to detect and destroy cancer cells. The main reason why your cancer risk increases as you age is because there are more chances for gene mutations to accumulate over time—and if your immune system isn't able to discover and attack those mutant cells, which often have signs of their mutant proteins on their surfaces, they can continue to grow unchecked. It's as if the gates to your body have some breaches, and your watchdogs took the day off: a double threat.

With weakening immunity on the horizon, it is extra important to do everything we can now to make ourselves stronger. That includes developing strategies to help us manage stress before it becomes a danger to our physical and emotional resilience.

UNDERSTANDING THE BENEFITS AND DANGERS OF INFLAMMATION

The health and resilience of your immune system determines your true age. The more your body is primed to protect you, the more likely you are to remain strong in the face of illness, and the faster you will recover if you do become ill.

All healing begins with a temporary flush of inflammation called acute inflammation. Seconds after you get physically hurt or your innate immune system is cued to a possible danger, blood vessels dilate so that more blood can get in, capillaries become more permeable so that proteins in your blood can go where they're needed, and white blood cells of the innate immune system move into the injured tissues to begin their repair work. Acute inflammation—a short burst of inflammation—keeps us healthy. That raised, red bump that appears after a mosquito bites you is beneficial. It is a sign of a healthy and functioning immune system, part of your body's response to help you heal. Acute inflammation is an intense reaction, and in a healthy system, the body calms that inflammation back down after it has served its purpose.

Sounds good, right? But when our bodies are constantly taxed by poor nutrition, not enough sleep, or a sedentary lifestyle; when we are subjected to too many external stressors and we don't have the tools to calm ourselves down, our immune system decides that we are in danger, and it prepares for battle. That means that the inflammation signal is turned to *on*—even when there is nothing to fight.

The result of all this battle preparedness is a low level of chronic inflammation, which accelerates cellular aging, or inflammaging. Your arteries, your joints, and your organs, are all susceptible to the damage of long-term inflammation. Chest pain (your heart), shortness of breath (your lungs), high blood pressure (your kidneys), blurred vision (your eyes), achiness (your muscles), pain and swelling (your joints), and rash or headaches (your blood vessels) are just some of the side effects of inflammaging. Chronic, systemic inflammation contributes to most age-related illnesses such as atherosclerosis, inflammatory bowel disease, rheumatoid arthritis, cardiovascular disease, cancer, osteoporosis and sarcopenia, Alzheimer's disease, neurodegeneration, and dementia, and as you know, researchers are also investigating the role of inflammation in the menopause transition.

BIOMARKERS FOR INFLAMMATION

Testing individuals for inflammation involves looking for certain biomarkers, or biological signs that can be measured and assessed to better understand systemic conditions in the body. Clinically, biomarkers usually refer to a molecule or protein that can be found in human bodily fluids, such as blood or saliva. Biomarkers are used by doctors to track the progression of a disease. For example, c-reactive protein (CRP) is a biomarker in the blood that is a measure of systemic inflammation. When you have the flu, your CRP levels rise. People with rheumatoid arthritis and other autoimmune diseases often have high levels of CRP.

Another biomarker for inflammation is interleukin-6 (IL6). Interleukin-6 has been linked to the age-related illnesses we've been discussing throughout this book, like osteoporosis, heart disease, and some cancers. Biomarkers are the holy grail for clinical trials. If you have a really accurate and responsive biomarker, like CRP, you can objectively measure the effect of a drug on a patient. Currently there are no consistent biomarkers for aging, although some have been proposed, like IL6.

If scientists could identify a biomarker that tracks with biological age, it could revolutionize the study of aging in humans. Remember, many aging studies rely on animals with short life spans so that researchers can observe the effects of proposed therapies rather quickly. If we knew of biomarkers that related directly to aging, we could test drugs for their effectiveness in humans and assess how effective they are without waiting twenty-five years to see if people are actually living longer as a result.

STRESS, IMMUNITY, AND THE
AGING MICROBIOME

Your microbiome kicks into gear at birth and changes steadily thereafter. Your first microbial assistants were some *Lactobacillus* bacteria that you got on your way through your mother's vagina and from her breast milk—and it was exactly the kind of bacteria you needed to digest her milk. For the next two years of life, your infant microbiome grew as your mother's contribution was joined by skin bacteria from all the people who held and kissed you, the food you ate, the dogs and cats and children that you played with. A stable microbiome is established during the first three years of life. By the time you became an adult, your microbiome will contain about 100 trillion bacterial cells from about one thousand different species. Microbiota are more similar among family members than

> By the time you became an adult, your microbiome will contain about 100 trillion bacterial cells from about one thousand different species.

people who are not related, but the place where you live and the people who live with you may be more important than genetic background.

The microbiome is dynamic and responsive—it is affected by everything around you and everything you ingest—and it evolves in response to your choices over the course of your life. As you get older, the health and diversity of your microbiome diminishes. With age, some people experience a weakened sense of taste or smell, have a hard time chewing, experience digestive issues, or have limited mobility, and all these factors can contribute to malnutrition, which may negatively affect your microbes. Prescription drug use is also common among the elderly, and the medications you take have a major influence on the population and diversity of your microbial residents. Furthermore, elderly people who live in a care facility tend to eat fewer fruits and vegetables and less fiber overall, dietary shifts that researchers have associated with less diversity in gut bacteria. With these shifts, more pathogenic and possibly dangerous bugs can move in, changing the balance of your bacterial population. A low-level, but chronic, gut immune response—in other words, inflammaging—is the result. The changes associated with an aging microbiome directly correlate with poor health and frailty in elderly adults. A less stable, less diverse population of gut bacteria is associated with an overall decline in health.

As with the rest of your cells, your microbiome is negatively affected by stress and inflammation and such factors can cause reduced losses in population and diversity. This loss in microbes is a loss for your immune system and can, in turn, create more inflammatory conditions in the body—exactly the kind of merry-go-round we want to avoid. The good news is, you can help maintain the health of your microbiome by eating a diverse and nutritious diet as you age that includes plenty of fiber. Healthy microbes, healthy body. And as it turns out—healthy mood, too.

THE POWER OF FECAL TRANSPLANTS

Half a million people in the United States suffer each year from a painful, horrible infection known as Clostridium difficile, which is an overgrowth of unhealthy bacteria that release toxins that attack the lining of the intestines. Some people afflicted with C. difficile are now opting to undergo a cutting-edge procedure: fecal transplants. (That's right, poop transplants.)

New research has shown that when healthy human feces—complete with healthy and beneficial microbes—is transplanted into C. difficile sufferers, something amazing happens: they get better. It's believed that the "good" bacteria in the healthy feces help kill off the "bad" bacteria in the C. difficile sufferers, allowing them to recover. More research is underway, but fecal transplants may offer hope for other difficult-to-treat infections in the future.

YOUR MICROBIOME AND YOUR MOOD

As we get older, changes in one system—like our microbiome—have an impact on other systems, including our psychological well-being. Once you learn about your microbes, you may be inclined to pay more attention to your "gut feelings" or your "gut intuition." That's because the microbes in your gut

Microbiome research is still evolving, but it looks like the same behaviors that we know prime us for unhealthy aging—eating processed foods, being sedentary, smoking, or being chronically stressed-out—may also decrease diversity in our microbiome. Eating a variety of healthful foods, exercising, and quitting smoking can help your microbiome age well. Here are a few more ways to give your microbial friends the support they need:

TAKE PROBIOTICS

Probiotics that contain *Bifidobacterium* and *Lactobacillus*, two genera of bacteria often depleted from our microbiomes with age, can help minimize the impact of taking antibiotics or not getting the proper nutrition. Probiotics like *Lactobacillus* have also been associated with improved mood and resilience in the face of stress.

EAT PREBIOTICS

Fruits, vegetables, and complex carbohydrates are full of prebiotics, like inulin, found in fiber. Prebiotics give the bacteria that make up your microbiome the food they need to thrive. When you take probiotics, you add bacteria to your intestinal mix. When you eat prebiotics, they ferment in your gut, creating an environment in which a healthy and diverse microbiome can thrive. So eat your asparagus, bananas, oatmeal, and legumes, all of which are high in prebiotics. Prebiotics are food for probiotics!

AVOID EXCESSIVE ANTIBIOTIC USE

As we discussed in Chapter 2, antibiotics extended human life span and are responsible for saving millions of lives every year. But regular use of antibiotics comes at a cost. Broad-spectrum antibiotics are just that: they don't discriminate between friend and foe, they just attack all bacteria, wiping out your microbiome in the process. Antibiotic use should always be accompanied by an increase in probiotics, good nutrition, and exercise.

LIMIT PROCESSED FOOD

Processed and packaged foods use antimicrobial preservatives to kill bacteria that cause food to spoil. These antimicrobial compounds work like antibiotics: they kill the bad microbes in the food (a good thing for food manufacturers who want their food to have a shelf life), but once ingested they can also negatively affect the good microbes within your gut.

MANAGE STRESS

Stress increases inflammation and diminishes microbial populations. In addition to killing off a lot of the good bacteria, stress can also create conditions in which populations of dangerous microbes flourish. Managing stress reduces inflammation and supports a diverse and stable microbiome.

actually have a direct line of communication to your brain. The intestinal tract contains a sophisticated neural network—called the enteric nervous system, or ENS—that is lined with more than one hundred million neurons. Some scientists even refer to the ENS as the "second brain." Not only does the ENS manage the basic functions of digestion, it also talks to your brain—and your brain talks back. This communication channel, the brain-gut axis, is

AN INTERNATIONAL BUFFET OF PROBIOTICS

For millennia, cultures around the world have been pickling and fermenting foods to make delicious treats that help with digestion. No one knew what the microbiome was a thousand years ago, but today we know that eating fermented foods is one of the best (and tastiest) ways to support your microbiome. Here are a few options to consider:

SAUERKRAUT
German, Austrian, and other eastern European cultures have long been fermenting cabbage to create this delicious dish, which is excellent in sandwiches and salads, or paired with almost anything you can imagine.

KIMCHI
Koreans ferment cabbage, carrot, and daikon with salt and peppers in a clay pot underground to make this spicy condiment.

YOGURT
Greek, Russian, Polish, and many Middle Eastern cultures have been fermenting animal milk—from cows, goats, or sheep—for thousands of years. Today, of course, you can find nondairy versions too, like coconut milk, soy milk, and rice milk yogurts, which are treated with *Lactobacillus* and other bacteria to make a thick, delicious, and protein-rich food.

KOMBUCHA
Russian, Chinese, and Japanese cultures have different versions of this effervescent fermented tea that's become popular in the United States in recent years.

KEFIR
Russians love this thick, sour beverage that's made from cow, goat, or sheep's milk treated with bacterial grains.

facilitated by the vagus nerve, which carries messages from your gut, including its microbes, to your brain and back.

The vagus nerve originates in the cerebellum and brain stem, then reaches down, caresses your heart, and threads through your abdomen. Your brain uses the vagus nerve to send information, via hormones, that can calm you down or rev you up, or tell your body when it's time to digest or rest. Your vagus nerve spends about 85 percent of its energy sending information about your gut to your brain. And that information isn't limited to the details of what you've been eating—it also includes emotional reactions. Feelings like fear and anxiety can originate in the gut and travel to the brain. Recent evidence from animal studies has shown that the overall health of the microbiome has an impact on behavior and mood. Since some of your moods and feelings originate with your microbiome, how healthy and balanced your microbes are affects your emotional health as well as your physical health.

The microbiome can even play a role in stress-related diseases of the central nervous system, like depression and anxiety. In the future, we may see therapies for mood disorders that have a microbiome-modifying factor. In fact, some emerging therapies for depression use electrical stimulation of the vagus nerve to trigger changes in the brain.

MANAGING THE EFFECTS OF STRESS

Stress begins with an initiating stimulus or event termed a stressor. A stressor is just a trigger—it can be something as simple as a raised voice or a late train. Over the course of a lifetime, as we move in and out of days and weeks and months and years, we encounter multiple stressors, some of which are a lot more worrisome than others: relationships, financial responsibilities, illness, loss. Stress is the response that we feel, within our minds and our bodies, to these difficult situations.

The effect of stress on our telomeres, deep within our cells, is something you are already familiar with from Chapter 6. The length of your telomeres is one of the many indicators of biological age. As you know, shorter telomeres are associated with chronic diseases and a shorter life. In every person, telomeres have a critical length—a point at which the cell is in danger. Once your telomeres reach this critically short length, the cell usually does one of two things. It either dies (via apoptosis) or it goes into senescence. The accumulation of senescent cells is a driver of aging and illness, because senescent cells create inflammation.

It's too easy to say "stress less." As we all know, stressors abound. We can't erase the guy driving too slow in front of us (my personal demon), or the automated "woman" from the cable company who just won't understand what we need. We can't make the neighbors stop yelling so loudly. We can't make the guy on his cell phone in the train seat next to us yell a little more quietly. What we can do is shift the way we respond to these cues. We can give ourselves some space—every day—to find a bit of quiet in a noisy world.

Many relaxation techniques, like massage and acupuncture, can help manage stress. But modalities that incorporate the mind as well as the body are usually shown to have the most long-lasting results. Meditation is a very

SIMPLE TECHNIQUES TO RELIEVE STRESS

- Take a walk. Spending time in nature can reduce high blood pressure and muscle tension as well as decrease levels of stress hormones.
- Take a seat. Meditation increases a sense of well-being that is associated with more telomerase activity in immune cells, which is associated with greater health and longevity.
- Try some downward dog. One study of 200 breast cancer survivors showed that after three months of regular yoga practice, the women had lower inflammation levels and higher levels of vitality compared with a control group. Other studies have found that people who practice yoga frequently and regularly have a lower inflammatory response to stress than people who have just started yoga; that yoga lowered inflammation in patients with heart failure; and that yoga improved glucose and insulin levels in diabetes patients.
- Let someone else work the stress out. Massage has been found to help with anxiety, depression, and physical pain, along with stress.
- Breathe. Aromatherapy, specifically using the scent of lavender, has been shown to lower stress levels and reduce anxiety.
- Laugh. Laughing relieves stress, relieves pain, and supports the immune system.

effective stress reducer whose benefits are both immediate and long-lasting. Scientists like Dr. Elissa Epel and Dr. Elizabeth Blackburn have conducted studies that offer evidence suggesting that meditation might have a beneficial impact at the cellular level by lengthening your telomeres. When you engage in what is known as a "mindfulness space intervention"—like meditation—you increase psychological well-being. Those feelings of well-being give you a momentary pleasure, and your cells like it too—meditating can raise your levels of telomerase within your immune cells. And more telomerase means longer telomeres and, in turn, healthier immune cell division—a domino effect that may promote longevity.

WHY I MEDITATE

Have you ever meditated? There are so many different kinds of meditation practices out there. There's guided meditation, Zen Buddhist meditation, mindfulness meditation, Hindu meditation, and countless others, all of which have their individual styles and benefits. If you're new to meditation, it's important to find a style that works for you so that you can adopt it as a "practice," meaning something that you will do regularly. The ability to sustain a long-term practice of meditation yields so many incredible benefits. Over the years, I tried many different kinds of meditation before I found a style that was the right fit for me.

Every form of meditation is practiced a little bit differently. Some focus on engaging with your breath while sitting in a position that you can maintain for the length of the meditation. Many schools of meditation practice the observation of the mind from a distance, with the idea that you let go of mental chatter and create a quiet mental space. Other forms of meditation suggest that you focus on a specific thought or a mantra.

For a long time I wanted to meditate, but I felt that I could never let go of my thoughts fully. So when people asked if I meditated, I found myself saying things like "I wish I could meditate . . . I've tried, but I'm just not good at it."

Then one day a girlfriend of mine introduced me to a different kind of meditation, called Transcendental Meditation, also known as TM. For me, it was the perfect fit. TM is transformative, and it's so easy to learn. Not just the easiest meditation style I've encountered, but one of the easiest things I've *ever* learned.

I try to meditate for twenty minutes a day, but on the days when I don't have twenty minutes to spare, I squeeze in what I can whenever and wherever I can. On the busiest days, sometimes I'll meditate for five minutes in the back of a cab.

In the TM tradition, each practitioner gets his or her own special mantra. You are given this mantra by your instructor, and it is just for you, not to be shared with anyone, not even husbands or sisters or best friends. I find it very empowering to hold this one innocent secret all to myself.

I begin my meditation by noticing my breath, and at the same time I start my mantra, which will come and go in and out of my thoughts while I meditate. Then the day rushes in and fills my mind, hopes and plans and stresses of the past, present, and future. But I don't feel pressure to push those thoughts out. Instead, my thoughts and my mantra dip and swirl together, engaging in a delicate dance until eventually they both settle down into the quietest part of myself. And soon there is just silence and stillness. I feel like I'm neither awake nor asleep, and my body feels as though it is immovable. It feels as though I've found a passageway into the deepest parts of myself, and that is when I realize that I am in the midst of healing myself from the inside out.

Meditation brings me back to *me*. I can actually feel the pieces of myself that have been broken off and scattered by all the stressors of life start to gel piece by piece, as if I'm watching a video in reverse of a glass shattering, shards scattered in midair brought back together. That's what meditation does for me: it brings me back to my whole self.

When I first learned how to meditate, I was all in. Isn't that how it always is when you try something new? For a while, I was completely dedicated—I meditated every day in the morning and in the afternoon. It was incredible. Everything in my life was benefitting from the calmness, clarity, and focus that I had gained from meditating consistently. And then, after about a year, I found myself starting to skip my daily practice, making the time for it only in those really dire moments when I *needed* it. Like when I missed my monologue that hot day in the Los Angeles Zoo, and meditation set me back on track. Experiences like that gave me another revelation: I had been given this amazing tool and I wasn't using it to its full effect.

Just like working out once or twice doesn't make you perfectly physically fit, meditating once or twice doesn't give you the full measure of its benefits. So I decided to start meditating at least once a day. Ever since I've done that, everything has changed for the better.

Meditation brings me back to *me*. I can actually feel the pieces of myself that have been broken off and scattered by all the stressors of life start to gel piece by piece, as if I'm watching a video in reverse of a glass shattering.

In my twenties I learned how much of an impact nutrition and fitness had on my body, and learning to use those tools changed my life. In my late thirties and early forties, this practice of sitting, used alongside the tools of eating well and moving often, has become the best gift I have ever given to myself. This tool is something that no one can take from me. It is always with me, and I can use it anytime, anywhere. It's given me a new kind of balance. It's given me a place I can go that is safe and quiet. A place within myself.

SUPPORTING YOUR WHOLE SELF PROTECTS YOUR WHOLE HEALTH

An organism's survival is directly tied to its ability to manage stress at the cellular level. Our cellular adaptability and resilience reflects the stresses accumulated over a lifetime: from the stressors of our childhoods to those of our adult lives. When the stress of adulthood is unrelenting—for instance, the chronic stress of long-term caretaking—our cells and our immune system are negatively impacted. The more nutritiously we eat, the more active we are, the better quality rest we get, the more techniques we develop to manage the inevitable stressors of life—the stronger our whole bodies will be. And the stronger our immune systems will be, allowing us to respond to illness or injury with the temporary bursts of inflammation that heal us instead of with the prolonged, chronic inflammation that hurts us.

The thing about meditation, about reducing stress, is that it's not only beneficial for your health, it also makes life a lot more enjoyable. You can relax, you can remember to be in the moment, to have a little bit more patience for your work or your family or your friends. Reducing stress protects your cells. It renews your energy. It makes your brain fire with more power, so you can focus on what you need to focus on and do what you need to do.

Learning how to peel away from stress allows you to feel younger and more alive and more at peace, even when the car in front of you drives too slowly, even when work isn't going the way you wish it would, even when you spy another silvery, gossamer strand.

CHAPTER 11

HOW TO PROGRAM YOUR SUPERCOMPUTER

Building a Stronger Brain

MY JOB REQUIRES ME to constantly reinvent myself. And I don't mean the hair and makeup and lines. I mean the preparation I have to do for every role. Each film is totally different from the one that came before it, and for me, that's the best part of my job: showing up every day and having to figure out how to do something I've never done before. Every scene is new, every shot is new—it's all a big puzzle that I get to be a part of solving.

I love getting the opportunity to learn new skills, like when I spent months studying martial arts for my role in *Charlie's Angels*. Or practicing singing for my role in *Annie*. Or brushing up on some old skills, like I did with my stunt driving for *Knight and Day*. The cool thing is that every time I learn something new, or improve or expand an existing skill set, I'm not just adding to my résumé. Every time I learn a new way of being physical in the world, like how to do a backflip hanging from wires, I'm not just building muscles—I'm creating connections in my brain.

Everything that I learn and experience today, like this language of science that I have been studying for the past few years, keeps building my brain. And the older I get, the more valuable these connections become. The same goes for you, because everything and everyone we love, you and me, everything we know and recognize, lives in the brain. Your brain holds your personality, your desires, your excitement. All of that, and so much more, is rooted in the connections between your neurons and the neurotransmitters that pulse information and sensation and awareness all over your body, as well as in the physical anatomy of your brain.

What makes you feel most alive? Laughing? Dancing? The feel of fresh air on your face? All that enjoyment originates in the four lobes of your brain. The parietal lobes are essential for skin sensation and also allow you to enjoy the warmth of the sun or the feel of six hundred–thread count sheets. Injury to the parietal lobes can result in confusion about space, an inability to recognize things by touch, or a general confusion about simple tasks like getting dressed.

The temporal lobes, behind your temples, hold memories of words, music, places, and tastes, and process sound. Remember your first concert? Your first kiss? The scent of your mother's perfume? Those connections live in your temporal lobes. The sunrise that you can see comes from the action of your occipital lobes, at the back of your brain, which take the visual information from your eyes via your retina and construct your conscious perception of the world. Damage to your eyes can make you blind, and so can damage to your occipital lobe.

Each of your brain's four lobes has its own set of responsibilities, and they all communicate with one another constantly and seamlessly while you live and work and love and observe the beautiful moon hanging above you. If you were to turn and tell the person next to you just how beautiful the moon is that your occipital lobes are helping you see, and how seeing that moon makes you feel, you'd be engaging your frontal lobes, behind your forehead, which govern organization, emotional expression, motor movements, and motivation.

All your most primal urges, from feeling hungry to wanting to kiss someone on the mouth, live in your brain. And yet the brain is also logical and calculating. It has special cells that triangulate space around you so you can find your way home. It has cells that hold your memories and the faces of the people you know. And it can run on autopilot, managing the support systems that keep you alive while you barely realize it, controlling your movements, your heartbeat, your balance, your breath.

THE MOST POWERFUL COMPUTER YOU OWN

Your brain boasts more processing power than any machine. In 2013, scientists used a supercomputer with more than eighty-two thousand processors to replicate a tiny portion of the human brain's neural networks. It took them forty minutes to replicate one second of human brain activity of just 1 percent of the brain. Forget your phone and your laptop and your tablet. The most powerful processor you possess is in your own body.

HOW YOUR BRAIN GROWS

The brain is a gray and lumpy mass that weighs about three pounds. It looks like the bark of a tree, and that is where its name comes from: when you look at a brain, the outer gray surface that you see is the cerebral cortex, and "cortex" is Latin for "bark." If you were able to smooth it out and unfurl all its many ridges and folds, your brain's surface area would be so large it could blanket a good-sized living room. Down the middle is a deep split that is full of neurons, or brain cells. Those neurons send messages back and forth between the left and right hemispheres.

What makes a brain healthy? There's the anatomy of the brain—the physical object. The mass of the brain is composed of neurons. While muscle cells contract and heart cells beat, neurons do something even more extraordinary: they reach out to each other, over and over again, creating hundreds of trillions of connections, or synapses.

The neurons that make up your brain have three parts: cell bodies, which hold a nucleus that contains your genetic information; dendrites, which are threadlike extensions that resemble the branches of a tree; and axons, the cell's roots, which are made of nerve fibers that burrow between spaces—synapses—to communicate with other cells. Those webs of connection become your neuronal networks, and they are the communication channels of a healthy and functioning brain.

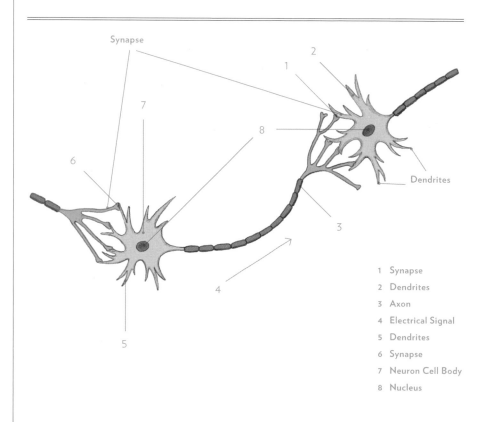

1 Synapse
2 Dendrites
3 Axon
4 Electrical Signal
5 Dendrites
6 Synapse
7 Neuron Cell Body
8 Nucleus

If you dissect a real human brain or spinal cord, the gray matter visible to the human eye is made up of the cell bodies and dendrites, along with their own supporting cells, called glial cells. The white matter is made up of axons covered in myelin sheath, a protective layer of fat.

There are as about 125 trillion synapses in the human brain—as many as there are stars in 1,500 Milky Way galaxies. As you live and grow and develop emotionally, or train and exercise and build your sense of balance, or engage in mindfulness activities like meditation, you build new connections and networks—and those connections will help you thrive even as the brain experiences natural losses with age.

Those connections are also what made your brain grow from an infant brain into an adult brain. When you were born you had a tiny little brain, only one-third the size of the one you have now. But it grew fast—to more than half your adult brain in just three months. The initial growth of your brain was spurred by your responses to new environmental stimuli, including interactions you had with the people around you. By the age of two, your brain had already begun to develop ways of thinking about and understanding people and social relationships. According to Matthew D. Lieberman, the author of *Social*, this ability in human young is unparalleled even among the adults of other species. This social outreach is the basis of why humans gather in groups, allowing us to create communities and societies and cities.

Throughout adolescence, the human brain experiences continued growth and change. By the time we're eleven, our sense of self is developing and emerging, shaped in large part by the social influences and belief systems around us. Our brains continue to grow throughout our teenage years and into our early adult years. In fact, humans take longer to mature than any other mammal: the brain doesn't fully mature physically until you're around twenty-five.

Research shows that teenagers use different parts of their brain to process emotions than adults do. Adult brain scans reveal more activity in the frontal lobes—the so-called executive part of the brain, involved in planning and impulse control—as compared with the scans of teenagers, who show more activity in the amygdala, which houses mechanisms for making impulsive, reactionary decisions. As teenagers grow into adults, the activity in their brains becomes more concentrated in their frontal lobes—which are among the last areas of the brain to mature.

The origins of modern neuroscience lay with our initial understanding of the connection between the frontal lobe and human maturity. In the mid-nineteenth century, a sweet young guy named Phineas Gage took a job he probably regretted later. He was a railway foreman, and as the railway was in the process of being built, things were still a little messy. Phineas just happened to be in the wrong place at the wrong time when a freak explosion shot a steel rod into the air, striking him in the head and piercing his skull. Miraculously, doctors were able to save him. In doing so, Phineas got a second chance at life, and neuroscience got its very first patient. Because while Phineas survived the blast, his personality did not. Before the accident, he was an amiable and responsible guy whom everyone loved to be around. Afterward, he was reportedly unreliable and impolite—basically, those close to him thought, a different man.

As it turns out, Gage's injury affected his frontal lobe, which is involved in functions like mood and social interaction. Gage's violent personality shift clued in doctors to the realization that some brain functions must be localized. The awareness that personality can be housed in the brain was the revelation that gave the burgeoning field of neuroscience its start. Since then, great advances have been made in technology—like fMRIs and CT scans—that have allowed neurologists to gaze more deeply into the brain than ever before. Sophisticated imaging tools allow us to see brain patterns and activities at the cellular level, revealing the profound effects of your neural networks on your overall function as a human being. These same tools give us the power to understand the effects of aging on the brain—how the brain can grow and connect, and how it can deteriorate.

Your brain cells have a longer life span than any other cells in your body. While your skin cells perish every three weeks, and your stomach lining cells last just three days, most of your brain cells have been with you since the day you were born.

Who we are is partially a product of what we learn and what we remember, but it is also based in our anatomy, which is why Phineas Gage seemed so changed to his friends and family after his accident. In a way, Gage's injury made him regress to an adolescent state, because he lost part of his "adult" brain.

An adult brain contains webs of connections and neural networks that are engaged in constant conversation across lobes and regions. These advanced networks are able to share information and observations and feelings and reasoning. That's one of the gifts of age and experience: the ability to deal with our emotions and make stronger, more rational decisions.

A TOUR OF YOUR NEURONAL UNIVERSE

Your brain cells have a longer life span than any other cells in your body. While your skin cells perish every three weeks, and your stomach lining cells last just three days, most of your brain cells have been with you since the day you were born. The brain contains about 100 billion neurons. For most of history, scientists believed that the brain couldn't create any new neurons, but in the last couple of decades they have discovered that the adult brain can generate some new neurons in very specific parts of itself, including the hippocampus (our memory center). This process is called "adult neurogenesis," and it means that the human brain is more plastic and capable of regeneration than we realized.

Over time, we continuously lose brain cells due to injury, illness, and the natural process of aging. Your neurons don't rely on cellular division to keep going; they rely on your body's protective mechanisms, like apoptosis, to safely clear away cellular debris and keep them healthy and active.

In addition to neurons, your brain also has specialized cells, called glial cells, that support your neurons by delivering them nutrients, insulating them, and holding them in place, among other duties. You have around ten to fifty times the amount of glial cells as you do neurons.

Like a computer and like your heart, your brain runs on electricity. Neurons have the ability to communicate information through the electrical and chemical signals of hormones and neurotransmitters. Your many thousands of dendrites receive information from other neurons, while your rootlike axons send information, not only to other neurons but to all kinds of cells, including muscle cells and organ cells. The cell body acts like a computer, taking in all the information it receives and processing it so that a decision can be made about what to do next. Some of these decisions involve your participation—do you calm down and continue the frustrating conversation you're having, or do you stop talking and slam the door on your way out?—and other decisions are made for you, like the neurons that keep you breathing.

An infant brain is small. An adult brain is larger. But the differences between the two go deeper than size: it's about the wiring. Brain imaging studies reveal that an infant brain has few connections as well as less white matter compared with the maturation of the adult brain, which is crisscrossed with pathways and strengthened by myelin sheaths. How many neural connections we have as an adult and how strong they are depends on how much learning and how many new experiences and conditions we have been exposed to over the course of our lives.

We received a tutorial on the way the brain works from Dr. Shennan Weiss, Assistant Professor of Neurology and Neuroscience at Thomas Jefferson University, who explained to us how the wiring of the brain is crucial to growth and learning throughout our lives. Scientists have long known that the brain builds new connections as it grows, but the fact that we can continue to build new connections and networks throughout our lives has only been established recently. Over the past forty years, researchers have discovered that the adult brain—which we once believed to be "hardwired" in its perceptions and responses—can, in fact, continue to forge new connections well into old age, changing neural pathways and creating new possibilities for how we experience and respond to everything from sensory stimuli to life events. This ability of the brain to "rewire" is known as neuroplasticity, and it means that actually, old dogs *can* learn new tricks. And that learning helps your brain stay adaptable as you get older.

When you learn something new, a neural pattern—a sort of microcircuit that connects various neurons, and sometimes various regions of the brain—is created in response. Each new experience that you have affects the connections between your neurons. Repeat experiences deepen those connections. Over time, neural networks are formed and become associated with anything connected to a particular experience. All the knowledge that feels second nature to you today started off as fledgling neuronal connections that, over time, grew into deeply ingrained networks.

This pool of experiences saved in your brain is known as your "cognitive reserve" or your "cognitive resources." Those resources were built up over the course of your whole life, and are influenced by everything from the education you've received to the languages you've learned to the emotions and life events you've experienced. Neuroplasticity allows us to strengthen our brains and make them more resilient against the threats of aging. Just as we build stronger muscles by challenging our bodies in ways that make them respond and grow, we can build stronger brains by challenging our intellect and helping our brains grow new connections.

In 2014, a thirty-two-year-old woman named Danielle Bassett became the youngest person to ever win a MacArthur Genius Grant for her insights into how the brain reconfigures over time. Bassett's work suggested that our brains are networking systems that grow according to the types of connections we make. The more your brain can make connections between existing hubs of information, the more it can figure out how to use what you already know to perform new tasks—and the more easily you can learn. It's an ideal way for the brain to communicate across its vast networks because it mini-

Just as we build stronger muscles by challenging our bodies in ways that make them respond and grow, we can build stronger brains by challenging our intellect and helping our brains grow new connections.

mizes the number of "jumps" it needs to make from one network to another. Connections between hubs allow information to flow freely across networks. The more connections you have, the more easily your brain is able to problem solve, to analyze, to remember. Learning builds connections—and that's exactly what you want, since, as you age, some hubs will naturally degrade.

Imagine that you need to fly across the country. If you miss your flight at a major airport like JFK in New York City, you can be on another plane to Los Angeles LAX within an hour. But what if your originating flight is at a regional airport that gets very little traffic? If you miss your flight, you're probably going to have to take an indirect route with lots of layovers, adding extra time to your trip. Building strong networks over the course of our lives creates neuronal hubs with more brain connections, so the hubs look more like JFK than a helicopter field in Maine.

Though around half a billion dollars is invested by the federal government into Alzheimer's research each year—a tiny fraction of the more than $200 billion that Alzheimer's costs the nation yearly—scientists still don't know exactly what causes it.

Just as brain connections can be made, they can also be lost. Diseases that break down cellular activity in the brain's information hubs contribute to cognitive decline in the elderly. It's thought that this is how Alzheimer's functions—by targeting the most important hubs for information sharing. The physical changes that come along with Alzheimer's show up as memory loss and confusion. When hubs are corrupted, as occurs with Alzheimer's, information can't pass through as readily as it once did, and people forget where they've put their keys, forget where they live, and eventually forget the things they've always known: their family members' faces, how to dress, how to speak. The more connections that exist among the various regions of your brain, the more you will be able to rely on alternate pathways, and the more your brain is able to exchange, share, and recall information when you need it. The stronger your networks are, the more resilient your brain will be in the face of degeneration.

The billions of neurons in your brain form trillions of connections at meeting points called synapses. When a neuron needs to transmit a message to another neuron, it emits an electrical impulse that prompts the release of a chemical messenger known as a neurotransmitter. The neurotransmitter fills the synapse, and the neuron on the receiving end reads the message it contains and emits its own electrical impulse in response. Neurotransmitters act as text messages between neurons; some signal a neuron to get moving and do its job (the "excitatory" neurotransmitters), others signal a neuron to stay put (the "inhibitory" neurotransmitters).

There are many different types of neurotransmitters, and they all have different jobs, sending information between neurons about your vision, your hormones, your moods, and just about everything else. These messages allow your body to know what to do, and when to do it.

Any number of factors can affect this messaging system, from nutrition to medications to the aging process. And while the impact of aging on the brain is unique to every individual, studies have found that most people experience changes in the following neurotransmitters as they age: glutamate, acetylcholine, gamma-aminobutyric acid (GABA), serotonin, and dopamine. It's important to note that these changes don't necessarily affect the brain unilaterally; they can be specific to one area of the brain, or even one type of cell.

GLUTAMATE is one of the amino acids that your body uses to make proteins. It is also the main excitatory neurotransmitter in the brain. Normal brain functions like emotional processing, motor behavior, and memory rely on glutamate.

Many studies in rodents have shown that glutamate levels decline with age, as do the densities of glutamate receptors in some brain regions. And brain imaging studies have detected decreased glutamate activity in aging human brains—particularly in the motor cortex, which controls your voluntary physical actions. Changes to the motor cortex may affect muscle strength and movement as we age.

ACETYLCHOLINE is an excitatory neurotransmitter. It makes muscles contract and it makes glands secrete hormones. Acetylcholine plays a critical role in cognition, particularly memory, and a decline in cortical acetylcholine function as we age has been associated with cognitive decline. When tested, patients with Alzheimer's disease have low numbers of acetylcholine receptors. The drugs currently used to treat the symptoms of Alzheimer's work to increase acetylcholine levels in the brain.

Continued

GAMMA-AMINOBUTYRIC ACID (GABA) is the major inhibitory neurotransmitter in the adult brain. Low levels of GABA have been connected to depression and anxiety.

Less GABA is produced as we age, but we can raise our GABA levels naturally through physical fitness, as one study demonstrated. Researchers had participants practice yoga three times a week for a period of three months, and monitored the subjects' GABA levels and mood. At the end of three weeks, not only did participants' GABA levels rise, but they also reported improved moods and lessened anxiety.

SEROTONIN has many functions: it makes your blood vessels constrict, helps you fall asleep, and helps regulate body temperature. But it is perhaps most well known as the "happy" chemical that makes you feel flush with joy and well-being.

Levels of serotonin neurotransmission decrease with aging, making it harder for our brains to "read" serotonin's message. Exposure to daylight and exercise are natural ways to enhance serotonin production, which is one reason why spending time outside in the summertime or getting in a good workout (or even better, going for a run or a hike outdoors) puts you in such a great mood. The happy effect of serotonin is important, because as we will soon discuss, your mood has an impact on your longevity.

DOPAMINE is an excitatory neurotransmitter involved in reward-motivated behaviors. From addictive activities like sex and gambling to addictive substances like sugar, alcohol, and drugs, the rush of dopamine released with them is what encodes these experiences in our minds as enjoyable or fun. But dopamine also has plenty of more wholly positive associations: it is the chemical released when we bond with others, and it has been linked to motivation and focus.

Dopamine levels decrease as we age, as do dopamine receptors. Parkinson's disease results from the loss of neurons that produce dopamine, so for patients with Parkinson's, the effects of dopamine loss in age are exacerbated. Studies have found that meditation can help to raise dopamine levels.

UNDERSTANDING DEMENTIA

Most cases of dementia or cognitive decline in the elderly are the product of a larger medical issue: Alzheimer's disease. Alzheimer's affects more than five million American adults and is the sixth-leading cause of death in the United States. The two major risk factors for Alzheimer's disease are age and gender: at age sixty-five and older, women have almost double the risk of Alzheimer's as men. Nearly two-thirds of Americans with Alzheimer's disease are women.

For many healthy adults, forgetting can be a natural part of getting older. Most of us will have days when we forget our grocery list, can't remember where we parked our car in the lot, or blank on the name of an old friend we happen to run into. As we get older, the ability to come up with the right words to describe what we want to say can also decline, and this type of decline is generally considered normal. It is not the same as dementia. Dementia is an illness, not normal aging.

The first symptoms of dementia often manifest as a difficulty remembering things for a prolonged period of time (not just forgetting a name on the spot), but they can quickly progress to more serious issues with planning, emotions, and self-control. Dementia makes it hard for the afflicted to maintain relationships, take care of their own finances, or manage a home. People with dementia may get lost in familiar places, ask the same questions repeatedly, behave erratically, or have trouble with their balance. Over time, their personalities can change, as can their eating habits and hygiene.

The two major risk factors for Alzheimer's disease are age and gender: at age sixty-five and older, women have almost double the risk of Alzheimer's as men. Nearly two-thirds of Americans with Alzheimer's disease are women.

Though around half a billion dollars is invested by the federal government into Alzheimer's research each year—a tiny fraction of the more than $200 billion that Alzheimer's costs the nation yearly—scientists still don't know exactly what causes it. At present, there are no diagnostic tests or definitive biomarkers for Alzheimer's, which means that we can't be certain if patients in clinical trials actually have Alzheimer's or another similarly presenting dementia. The slipperiness of the diagnosis may be one of the reasons why so many Alzheimer's drug trials have failed: researchers can't know if they are actually treating Alzheimer's patients with Alzheimer's drugs. While there are a number of drugs in use and development to lessen the symptoms of Alzheimer's, there is currently no cure or any drug that reverses the course of the disease.

That said, drugs may not be the only answer in the battle against Alzheimer's. We had the opportunity to speak with Dr. Dale Bredesen, an expert on aging and the brain, in his office at UCLA (he has a joint appointment with the Buck Institute as well). Dr. Bredesen and his team are spearheading research that examines the impact of nutrition, sleep, and fitness on brain health. He described to us a new study his team is working on that determines an extensive metabolic profile for each person, then uses a personalized combination of exercise, nutritional supplementation, stress relief, hormone modifications, and sleep optimization to treat patients in the early stages of Alzheimer's disease (he calls this the MEND protocol—Metabolic Enhancement for Neurodegeneration). While more clinical trials are needed, Dr. Bredesen has published promising evidence that this multimodal approach can help reverse cognitive decline and increase function in patients with early Alzheimer's disease.

PLAQUES AND TANGLES

When people with Alzheimer's disease pass away and their brains are autopsied, scientists observe two physiological changes: tangles within the cells and plaques outside the cells in parts of the brain like the hippocampus and the cortex. Tangles and plaques are made of proteins (tau and beta-amyloid, respectively), and these proteins are the hallmarks of the anatomical deterioration that causes the memory and personality shifts we associate with Alzheimer's disease.

Amyloid plaques and tau tangles are clusters of indigestible protein that seem to injure nearby healthy cells in the brain. We all have cells that secrete these proteins. The danger, as we age, is the potential for these proteins to hang around—instead of being flushed out as regularly as they once were—and stick together, causing a buildup.

Their shape is partially to blame. As you know from Chapter 5, proteins are

3-D structures made of tubes folded in and curled over themselves. That unique structure is what allows them to do their jobs. When proteins misfold and the sticky surfaces catch one another, a plaque of stuck-together proteins results.

While we all secrete these tau and beta-amyloid proteins, not everybody ends up with neurodegenerative diseases. Some genes can multiply your risk of developing a degenerative brain disease, but remember, your environment plays a role in the expression of your genotype. Scientists now know that exposure to new learning and experiences is one of the best ways to protect yourself against cognitive decline. So is getting adequate rest. While we sleep, our brain cells clean and repair themselves, getting rid of some of those sticky misfolded proteins in the process.

THE CARE AND FEEDING OF A HEALTHY BRAIN

When it comes to protecting your brain as you get older, making new connections matters—and your overall physical health as well as your emotional health play important roles too. Cognitive decline can be affected by a number of physical and emotional conditions, including family history, heart disease, high blood pressure, diabetes, poor eating habits, physical inactivity, low levels of education, stress, depression, and social isolation.

Studies have shown that being fit, eating nutritious foods, getting enough sleep, and reducing stress, as well as a history of education and current mental stimulation, are all associated with decreased risk for cognitive decline with age.

PROTECT YOUR BRAIN WITH FITNESS

Studies have demonstrated that obesity and metabolic disorders are associated with poor cognitive performance, cognitive decline, and dementia, while physical exercise has been associated with a decreased risk of age-related cognitive decline and Alzheimer's disease.

Increasingly, controlled studies are telling us that people who exercise three times a week or more are likely to experience dementia later in life than people who are sedentary. Exercise also helps you think more clearly, reduces inflammation and stress levels, and releases growth factors in your brain like BDNF, which aids in cellular health (see page 134 for more information). One study of people over 65 showed that people who regularly engaged in moderate-intensity exercise (like walking or jogging) a few times a week had an increased volume in the hippocampus—the part of the brain that governs memory and is particularly vulnerable in Alzheimer's compared to their sedentary peers. This was true even for people with a genetic predisposition to Alzheimer's disease. So be sure to take your memory out for a stroll every chance you get.

PROTECT YOUR BRAIN WITH SLEEP

Adequate rest is key to brain health. As we discussed in Chapter 9, sleep gives our cells the opportunity to carry out all the protective processes that keep us alive and well—including clearing out the harmful plaques that can accumulate in our brains with age. It is thought that buildup from those plaques (which is linked to Alzheimer's) can also lead to sleep disturbances as well. Sleep also helps protect us from stress and depression.

PROTECT YOUR BRAIN BY REDUCING STRESS

Stress affects our cognitive function, and as our brains become less resilient with age, stress begins to hurts us more and more. At any age, prolonged stress or anxiety can reduce brain volume. It can also hurt our ability to make new synaptic connections, which we know is important for maintaining neuroplasticity. When you're young, your brain bounces back when the stress abates. In middle age, it becomes harder—stress can shrink dendrites and synapses—but your brain can still recover. As we get even older, our brain is less able to recover what it has lost. Be sure to look out for older friends and family members who are dealing with the stress of loss or illness, especially at advanced ages.

PROTECT YOUR BRAIN WITH MEDITATION

Mindfulness-based activities like meditation and relaxation exercises have been linked to lower levels of the stress hormone cortisol. And a regular meditation practice has been shown to change the architecture of your brain in ways that are beneficial and protective.

In 2011, researchers documented that a group of participants who meditated for just thirty minutes a day over the course of eight weeks exhibited physical changes in gray matter. In the meditators they saw increased gray-matter density in the hippocampus, the brain's memory center, coupled with decreased density in the amygdala, the brain region involved in regulating stress and anxiety. Another study in 2012 showed that just two months of meditation resulted in reduced amygdala activity, indicating improved response to stress and emotional stability . None of the structural or activity changes were seen in the control groups, who did not meditate.

PROTECT YOUR BRAIN WITH LEARNING, LANGUAGE, TRAVEL, AND INTERESTS

New brain connections are forged by new learning and new experiences. A large study in India showed that being bilingual delayed the onset of dementia by 4.5 years compared with monolingual adults. In general, the more interests and activities we have in our lives, the more opportunity our brains have to make crucial connections. One study in France observed the habits and lifestyles of 2,400 people over the age of sixty-five and found that those who traveled, were knitters or gardeners, or had odd jobs were less likely to develop dementia than their peers who were less active.

PROTECT YOUR BRAIN WITH MUSIC

My husband is a musician, and he has an amazing collection of guitars that adorn our home. I get to listen to him as he strums on them, working on new songs or just playing for pleasure. It soothes me and keeps us both company. We also have a record player, and often there's something spinning and filling the air with sounds that make us feel nostalgic, romantic, or like having an impromptu dance party. Playing music, listening to it, creating it—surrounding ourselves with songs that resonate with our emotions—can affect not only the way we feel, but also the way we think and how our brains are wired.

Studies show that when people listen to music and when they play music, their entire brains light up with increased neural activity. That's because you engage the visual, auditory, and motor areas of the brain when you play music; the more you practice, the stronger these areas of the brain become. People who play music regularly can develop their corpus callosum, which is the connection between the two sides of the brain. A stronger, thicker corpus callosum facilitates more rapid signaling across the brain, making a musician's brain one that may have more agility for problem-solving.

Music can also help with memory in dementia patients. Musicians have an enhanced ability to create and store memories, and also to recall them when they need them.

MEMORY MAKER

For many adults, changes in cognition will be a part of healthy aging. For some, the changes to the delicate networks of connections and the physical anatomy of the brain will be more severe and may result in the symptoms we associate with brain degeneration, tremors, and memory loss.

There are things we can do now to minimize the severity of those changes in the future. We may think of the brain as being in charge of our life choices and experiences, but the truth is, it's a feedback loop, because our choices and experiences also shape our brain cells and our networks, and thus the very architecture of our brains. Our brains can become stronger and more resilient by doing many of the things we already love to do: learning, moving, reading, resting. The same activities that help us stay engaged, active, and mentally sharp in the short term are also the very actions that can protect our minds over time.

That's one more reason I want to keep hiking, keep meditating, keep learning, keep cooking, keep listening to music and exposing myself to new experiences—because I want a strong and healthy brain that I can rely on. I want to protect the memories I've already made—and I want to keep making beautiful new memories for as long as I possibly can.

CHAPTER 12

LOVE BIG

Celebrating the Joys of Connection

M Y BIRTHDAY IS IN the summer, and every summer since I turned forty, I have celebrated another year passing. Each birthday celebration has been special in its own way. There was a party that I hosted. I cooked all afternoon and fed the people I love (and then spent three hours doing dishes, accompanied by a couple of steadfast pals). There was a surprise party that my best friend threw for me, when my husband, when he was still my boyfriend, took me for dinner while all my friends waited in a bathroom for twenty minutes for us to arrive (and yes, I was surprised). Then there was the dinner party he hosted for me when I turned forty-three, an elegant little meal that I enjoyed not having to do the dishes for.

All these celebrations were different, with a single constant—that bolstering, bountiful feeling of being surrounded by friends, the people who make me feel loved and whom I love. At each party, I knew that my friends thought they were there to celebrate me. But really, I was celebrating them. I was celebrating all the joys we had shared, and would share. All the sorrows we had held for one another, and would hold. We are born into families, and then we go out into the world and we make strangers into our families. And these extended friend-families are with us for our whole lifetimes, sometimes.

That's one of the great gifts of life, isn't it? The opportunity to find the friends who will help shape us and shape our destinies.

I have some girlfriends whom I grew up with, whom I've known for decades; I have friends whom I've encountered in various ways and at various points along the journey that is life; and I have friends whom I've met only recently but feel like I've known forever. There are the friends I rarely get to see or talk to—but when we reconnect, it's like no time at all has passed. And then there are the friends who live nearby, who stop over to ask a favor, to borrow something, or to say hello over a cup of tea. There are the big occasions we celebrate together—we organize surprise showers, we dance at each other's weddings, we plan elaborate holiday dinners. And then there are the impromptu afternoon picnics, with baskets of fried chicken and kids playing together in the grass. All these relationships, all these occasions, are an essential part of my life.

Connections like these feel so good for our hearts and our spirits that it shouldn't come as much of a surprise to learn that they are actually good for our bodies too.

THE LINK BETWEEN HEALTH AND COMMUNITY

Human beings are social creatures—our brains are hardwired for socialization. These social connections offered valuable protection for our ancestors, who needed numbers to survive under harsh conditions. One person couldn't possibly do the hunting, keep the fire going, and fight off predators alone. Humans have always lived in groups and formed societies for their safety and well-being. A thousand years ago, a person completely on his or her own would never have made it to adulthood (and even in developed societies, most wouldn't be celebrating birthdays into their forties).

Today, even with all our creature comforts, even with advances in medicine and technology and longer life spans, connection is still essential to our survival. Because even though we can live alone, work individually, and order a pizza with the touch of a button, the number and quality of the human connections that we have plays a crucial factor in our health and well-being. This is especially true as we age, when isolation is linked to increased risk of mortality.

It may seem obvious, but it's an essential point for all of us to understand because in the United States today, people are becoming increasingly isolated from one another. In 1970, 17 percent of people lived alone, compared with 28 percent in 2011. According to the results of an AARP study published in 2010, more than 35 percent of adults aged forty-five and older reported feeling lonely. That same study also found that loneliness was a significant predictor of poor health.

The value of social connection is worth noting in a book about healthy aging because as we grow older, our likelihood of becoming isolated increases. As the years pass and our lives change, so do our relationships—old friends and neighbors move away, children grow up, responsibilities shift, jobs end. As we grow even older, health issues and changes in income can lead to increased isolation. And the impact is profound: studies on the effects of social isolation on the elderly have found that those without adequate social interaction were twice as likely to die prematurely.

It's a vicious cycle: illness compounds isolation, and isolation compounds illness. If we want to grow older with health, maintaining social connections is key.

We can't live on love alone. But we can't live without it, either.

LOVE IS A CIRCLE

You need social connection, and so do the elderly people in your community. The elderly are especially likely to suffer from social isolation and loneliness as friends, family members, and partners pass away or move to care facilities, and limited mobility impairs new social connections. So if you know any people who need visitors: visit them. It's good for their health. And it's good for yours too—studies have shown that people who volunteer live longer.

CONNECTION IS THE BEST MEDICINE

Isn't it great when science validates the things we feel intuitively to be true? In writing this book, we discovered time and again just how beautifully biology can explain how and why the stuff in life that feels so good really is so good. Friendship, happiness, joy; sharing laughs and sharing tears; these are not just the elements of a good life, they are the elements of a long and strong life. Friendship affects health. Happiness affects health.

Over the past few chapters, you've learned about the dangers of inflammation and inflammaging. About how neural connections influence cognitive health. About how aging takes place deep within your cells, where your telomeres are working to protect the delicate ends of your DNA strands. All these common causes and markers of aging—inflammation, cognitive decline, telomere length—have been tied to social activity, happiness, and community.

When we think about health, we tend to think about what we eat and how often we exercise—and we should, because good nutrition and fitness practices can reduce inflammation and help us age with health. But so can the quality of your relationships. In one study, 135 women between the ages of sixty-one and ninety-one answered questions about their quality of life.

> Friendship, happiness, joy; sharing laughs and sharing tears; these are not just the elements of a good life, they are the elements of a long and strong life. Friendship affects health. Happiness affects health.

Those who said they had many positive relationships and had a purpose in life exhibited lower levels of inflammatory markers than the women who had fewer social connections.

In another study, University of Michigan researchers who tested 3,610 people between the ages of twenty-four and ninety-six found that even ten minutes of social interaction improved cognitive performance. People with regular social ties were significantly less likely to demonstrate cognitive decline when compared with those who were lonely or isolated. And social connection has been associated with longer telomere length, which is associated with greater health and longevity.

The science proves it. More happiness and positive relationships mean less inflammation. More socializing means less cognitive decline. More love means longer telomeres. More laughter means a stronger immune system. That means that the most common dangers for rapid aging can be lessened by having friends. By loving people. By offering kindness and support to the people around you, and receiving the same in kind.

QUALITY OVER QUANTITY

You don't need a million friends, or a million followers. You need quality contact with quality people. Studies show that being part of a supportive community is good for your health—and the key word is "supportive." It's not just about having people around you. It's about being surrounded by people whom you value, who value you, and whom you feel valued by. We need to feel supported, or our risks for depression and illness can rise. The danger of having troubled relationships or too few relationships isn't just that of feeling a little lonely or misunderstood—the lack of healthy social connection is also tied to health risks as we age. In one study, married women who reported having stressful relationships with their husbands had more biological and psychosocial risk factors for cardiovascular disease than women in strong relationships.

Consider the results of a study that observed 6,500 men and women over the age of fifty-two to assess the impact of social connection on health as we age. Researchers took into consideration two types of isolation: social isolation and loneliness. Social isolation was determined by objective measurements: how often the subjects made contact with other people. Loneliness, however, was defined subjectively—after all, you can be surrounded by a roomful of people and still feel lonely.

The study revealed that people who were socially isolated had an elevated risk of developing high blood pressure, heart disease, infections, cogni-

We all get sad sometimes. We all feel frustrated, low, anxious. Most of the time, we will soon feel better—but sometimes we can fall into a pit of despair and find it very challenging to crawl out again.

Each year about 7 percent of US adults experience major depression. Women are 70 percent more likely than men to suffer from depression. For millennia, thinkers have been trying to decode depression, from fourth century monks describing it as a "visiting demon" to Sylvia Plath's famous depiction as being trapped within a "bell jar." Today, researchers have some insight into the root causes of depression and have devised therapies that can help people recover.

Depression can range from mild and temporary sadness to clinical or major depression. If you've suffered a loss, like the death of a loved one, the grief that results is natural and, usually, will pass with time. A clinical or major depression is a more serious psychological condition for which professional help is a must.

SYMPTOMS OF CLINICAL DEPRESSION INCLUDE:

• Persistently depressed mood over a two-week period (meaning every day)
• Significantly reduced ability to feel pleasure
• Changes in sleep
• Weight gain or loss
• Shifts in interests that affect normal daily activities
• Shifts in interests that affect participation in meaningful relationships

Like most complex human illnesses, depression is believed to be caused by a combination of genetic, biological, and environmental factors. While most people's happiness improves with age, there is also a rise in depression with age that is linked to illness, isolation, and loneliness. Shifts in physical health and changes in lifestyle can also have a negative effect on mood, decrease enjoyment, and increase depression. Older adults who have experienced one or more debilitating chronic diseases, or who are hospitalized or rely on home care, are at a higher risk of developing depression.

If you or someone you love is suffering from depression, please seek help from an expert. Treatments for depression include counseling, medication, and nutritional therapies. Meditation and exercise have also been shown to be beneficial for managing symptoms of depression.

tive degeneration, and inflammation, while loneliness was linked to higher risk of cardiovascular disease, high blood pressure, and a poor immune response to stress.

According to the study, there was no sex difference when it came to social isolation—women and men were equally as likely to lose contact with the people around them, especially when they were unmarried, had low incomes, or were in poor health. But loneliness—feeling isolated even if you are objectively in contact with people—was found to be most common among women.

If you find yourself feeling lonely, it's important to seek connection that is meaningful to you—or help. Loneliness compounds depression, and depression compounds loneliness. Women are also twice as likely as men to suffer from depression, especially during the menopause transition. If you find yourself feeling alone more often than not, call a friend you trust, seek counseling, visit a community center, or find a volunteer group. Even if it feels awkward or outside your comfort zone at first, you will thank yourself later.

TAKE SOME ADVICE—AND GIVE SOME

If you've ever been totally confused about something—and who among us hasn't?—you know that the experience of having someone sit down and patiently explain what you don't understand is one of best, most reassuring feelings there is. When we were working on this book, we experienced that sense of relief and connection more than once.

We had never met Dr. Richard J. Hodis before he invited us to lunch and talked to us for hours about human life expectancy, just as we had never met Dr. Judith Campisi, who introduced us to all of her colleagues, or Dr. Elissa Epel, who served us hummus while she made us feel better about the damage happening to our telomeres. All these very busy people opened their doors to us and shared their knowledge with us, happy to spend the time because they know how important it is for you, holding this book in your hands right now, to get a glimpse inside their research laboratories. They also understand better than any of us the importance of connection, and no matter how crazy their schedules may get, they make time for conversations.

That is how we build community—by opening up and by being open. Sometimes it's about seeking advice from an expert or a mentor; sometimes it's about being someone else's mentor or expert; and sometimes it's about listening to what the next generation has to say. Part of the wisdom that comes with age is the confidence to heed the counsel of somebody who is less experienced or younger than we are. I don't care if you are the CEO of a Fortune 500 company with all the experience in the world, I promise you that there is a junior assistant or an intern somewhere in your corporation who understands something that could benefit you. And the truth is that as we get older, our doctors, lawyers, and even bosses will be younger than us. Connection can come in many shapes and sizes and ages.

With so much that's unknown and unknowable ahead of me, I find it immensely comforting to know that I'm not alone. I have mentors and teachers to guide me. I have my family. I have my friends. And all of those connections will make me stronger, healthier, happier, and more fulfilled.

LIVING WELL STARTS WITH REALLY *LIVING*

I was recently out to dinner with a group of friends, and there was a table full of ladies nearby in their fifties and sixties. They were a rowdy group who started out with a bottle of wine and just kept laughing more and more loudly as the night progressed. They were having a great time. They didn't seem to notice anybody else in the room.

And my friends and I looked at one another and said, "That's how you do it! That is how it's *done*!" These women were positively glowing with the delight of being together. They weren't wondering about how they looked to anyone else in the restaurant. They were completely in the moment, just living it up for the evening. They didn't seem to be worried about how *old* they were. They seemed to be focused on how *alive* they were.

To me, that is one of the gifts of aging. The ability to be present. To be confident. To be fully, unapologetically, yourself.

The thing is, contrary to the messages that women receive every day, life isn't only for the young. Enjoying life isn't about being young. Beauty doesn't depend on looking young. Joy and companionship and connection and all the best parts of life don't only happen when you're young.

So if you want to age well, consider what makes you feel most alive. Consider what fills you with love and anticipation and wonder and joy—and then go and do some more of that, please. Because isn't that the point of all this self-care? To enjoy our lives. To enjoy our families and friends. To look forward to possibilities and improbabilities and the best surprises that life can bring.

The thing is, contrary to the messages that women receive every day, life isn't only for the young. Enjoying life isn't about being young. Beauty doesn't depend on looking young. Joy and companionship and connection and all the best parts of life don't only happen when you're young.

HAPPY MIDLIFE TO YOU

When someone asks you how old you are, what do you say? I know that there are plenty of women in my line of work—and many other fields in which young people rule the roost—who like to trim their age back a little bit because they believe it makes the people around them feel better when they say they are younger. We live in a culture that celebrates youth, and none of us are immune to the pressures of wanting to seem forever young. But one of the reasons I wrote this book is because I believe that we would all be a lot happier, feel a lot better, heave a big sigh of relief, if we could just answer "how old are you?" with the truth. Without fear. Without hesitation. Without shame.

Because I believe that age is a marker of achievement.

Shouldn't we be congratulated for all that we've accomplished over the decades instead of being asked to pretend that they didn't happen? Shouldn't those years count for something? I think they should. I think those years are a part of what make us the people we are today. I think we should get to keep them, all of them, and proudly. Every time I celebrate a birthday with people that I care about, I think about how blessed I am, how lucky I am, what a gift each year is. That's why I celebrate. Because I have lived long enough to have learned these lessons, to have earned these relationships, to have discovered all these new layers of myself that I didn't know existed years ago—or that hadn't yet developed.

Longevity is a gift we should all be celebrating. The more years you have enjoyed, the more time you have lived, the more chances you have been given. Chances to take on challenges, to explore possibility, to create your life's story. Time to forge meaningful connections with others—time to love deeply, to be hurt deeply, to become a role model for all the daughters and nieces and granddaughters and sisters and young friends in your orbit.

So I would like to propose a new way to think about getting older. I would like to boldly suggest that we take those years back. No hiding. No apologizing. No deleting. No erasing.

Instead of dreading midlife, instead of fearing forty or fifty or whatever the magic number is for you, decide now to honor it and to own it. Let's push the midlife crisis off a bridge and throw ourselves a party instead. The midlife celebration: a personal holiday that celebrates the journey we've made to get here, and the unexpected places we have yet to discover.

Because the best way to age healthfully is to live fully. To take care of your body and your spirit in this moment, where you are now. You can spend your energy on love and not on worry. You can love the world, you can love the people around you, and you can love the person you have spent all these years becoming.

Yourself.

URING THE FINAL DAYS of writing this book, we retreated from the city to have some space to think about what it all meant, all these facts we had been absorbing over the year about what aging is and how it will affect us and how we can ride this wave of change. On our way to the retreat, we drove north from Los Angeles in separate cars, taking different routes, winding up in the same place. And then we were there, sitting together with our computers and some snacks and looking out toward the water, a very beautiful and peaceful vista, the perfect space for contemplation. And Sandra looked up at me suddenly and asked, "How did we get here?"

I knew just what she meant, because after writing two books together we have developed a shorthand when we speak. She wasn't referring to the busy 405 cutting across the landscape, or the 1, curving between the Pacific and the Santa Monica Mountains, even though people in California often begin conversations by talking about freeways. She meant the long road, the whole journey, the real trip that we had been taking over the course of our lives. How did I get here? How did I go from a girl who grew up in a neighborhood in Long Beach situated between two freeways and a railroad track to a woman with a career that's taken me all over the world, from the unexpected turn that made me an actor to writing a book about aging? How did Sandra get here, three thousand miles away from where she grew up in Brooklyn, from a kid who would spend the weekend with a stack of novels to a grown woman who writes bestselling books?

Had you approached us when we were fifteen to tell us our life stories, neither of us would have believed you. There's no way to know up front where life will lead us. But at a certain point, you gain the opportunity to look back and find the thread that shows you where you have been, and where you might want to go—and where you are right now.

For years, psychologists have been investigating the idea that part of growing older is considering the greater meaning of our life's story. Erik Erikson, a noted developmental psychologist, believed that in our later years, our natural instinct is to make peace with our lives. At a more mature age, we have the ability to look back from a distance so that we can begin to put all the pieces together and understand the purpose of our journey. It's not a crisis; it's a rite of passage.

SELF-DEFINITION

When I started my career in Hollywood, I was a very young woman. It was more than twenty years ago, when my first movie, *The Mask*, came out, and I was pretty naïve. I was new to this business. I had never acted in a film before. I didn't know if there was a right way to handle the experience, and I certainly didn't know how to deal with the emotional and psychological components of it. So when the reviews came out, I read them. Many of them were very complimentary; people said some lovely things about me and my acting. And I thought, *Well that's nice*. It made me feel great. There I was, new to this big and scary world, and people were being so kind and welcoming. And so I kept reading, and then I saw some reviews that were not so complimentary. And those made me feel pretty bad, to tell you the truth.

So I had to take a step back and consider what I believed to be true. I knew I couldn't choose to believe only one set of reviews; if I believed in the glowing reviews, than I'd have to believe in the negative reviews—the ones that were kind of making me feel like crap.

I realized then that it was better to put all of it aside. I could not move forward in this line of work if I was going to rely solely on the opinions of others to define what was real for me. I had to let my experience of my work—not someone else's—define how I saw myself. Did I do the best I could do at that time? Did I succeed in what I wanted to accomplish? Did I hit the mark I set for myself? Or could I have done better, tried a little harder? Did I hold back from giving it my all? These are the standards that define success for me.

The valuable lesson that I learned back then is that looking outside yourself for approval is just plain dangerous. That's why, when I turned forty, and those journalists criticized me for being a human who gets older, I was ready. I was not a naïve girl anymore. I was a woman who knew that I was the only critic who mattered, and that it is my job to create the meaning in my life.

Just as it is your job to create the meaning in yours.

SELF-HELP

The foundations of a strong and healthy life are the same as the foundations of a beautiful and memorable life. Nutrition is about enjoying joyous meals with friends and family. Fitness is about being active, getting outside, discovering new abilities and new people and new pockets of the world. Getting rest is about letting go of little nagging cares so you can truly restore your body and mind for another day of living. Stress relief is about caring for your spirit and connecting to your innermost self. And being social is about connection, joy, laughter, sharing, loving, and learning.

What will the future hold? There's no way for any of us to know. Not one of us can anticipate what will happen over the next years. We know that it will be hard sometimes. We know that there will be challenges. We know that we will need resilience and determination. Everything else is a guess, a hope, a wish. Like the wish we all share to live lifetimes that are full of strength and love.

The stories of our lives twist and turn, have highs and lows, collapse and expand, falter and begin anew—and over time, a combination of hard work and plain luck lands us in places that we never could have imagined. Some of the twists will be more wonderful and rewarding than we ever could have dreamed. And some of the turns will prove more challenging and unpredictable than we ever wanted to consider possible. But even amid all of this uncertainty, there is one fact that we can rely on—and that is that the stronger we are when challenge taps us on the shoulder, the better our chances of surviving whatever comes our way.

So let's agree to stand together. To look forward, together. To the good times, to the hard times, to the unexpected times. We can't predict the future, but we can invite it. We can prepare for it. And we can give ourselves a chance to live longer, stronger.

EPILOGUE

PEGGY FREYDBURG WAS EIGHTY years old when she wrote her memoir, *Growing Up in Old Age*. She was ninety when she started writing poems. Over the next sixteen years, Peggy wrote eleven books of poetry, which she shared with her friends. When Peggy was 106, her friends decided that the rest of the world deserved to read her poems, too, and they gathered her selected works into *Poems from the Pond*. "Chorus of Cells" is one of the thirty-two poems in that collection.

We read "Chorus of Cells" aloud one afternoon in the desert, surrounded by nothing but sky and sand, and both of us fell silent afterward, caught in the magic of Peggy's words.

Peggy passed away at age 107. This epilogue is a tribute to her and to every person who embraces the notion that creativity is ageless and that it is always the right time to discover yourself.

"A Chorus of Cells"

by Peggy Freydburg

Every morning, even being very old (or perhaps because of it),
I like to make my bed.
In fact, the starting of each day
unhelplessly,
is the biggest thing I ever do.
I smooth away the dreams disclosed by tangled sheets,
I smack the dented pillow's revelations to oblivion,
I finish with the pattern of the spread exactly centered.
The night is won.

And now the day can open.
All this I like to do,
mastering the making of my bed with hands that trust beginnings. All this I
need to do,
directed by the silent message
of the luxury of my breathing.
And every night,
I like to fold the covers back,
and get in bed,
and live the dark, wise poetry of the night's dreaming,
dreading the extend of its improbabilities,
but surrendering to the truth it knows and I do not;
even though its technicolor cruelties,
or the music of its myths,
feels like someone else's experience,
not mine.
I know that I could no more cease
to want to make my bed each morning,
and fold the covers back at night,
than I could cease
to want to put one foot before the other.
Being very old and so because of it,
all this I am compelled to do,
day after day,
night after night,
directed by the silent message
of the constancy of my breathing,
that bears the news I am alive.

ACKNOWLEDGMENTS

IT TAKES A VILLAGE to write a book. So I'd like to thank my village for gathering around and giving me their wisdom, guidance, and love along this journey.

Julie Will . . .

I can't thank you enough for your passionate dedication to the quest to inform, the invaluable contributions you made to the content of this book, and the endless hours spent helping to shape and to ensure that we had the best book possible. You are more than an editor you are a collaborator and a superstar!

Jesse Lutz . . .

Without you the train would be off the tracks and the ship would be on the rocks. Your keen sense of story and your love for the written word helped us stay our course. . . . You are a true navigator. Without you I'd be lost. I love you, Frienders.

My Husband . . .

I never knew what love was until I found you. How did I ever live without your love, your friendship, your brilliant mind, your genius sense of humor, and your enormous, loving, beautiful heart? Thank you for being my best friend, my teacher, my partner in life. You inspire me every day with your courage and devotion. Thank you for all the support and encouragement in this book and in everything in life. I love you forever and ever.

My Mommy . . .

I'm so glad I'm a Mama's Girl. Your love is my super power. It's because of you that I have any courage at all. I love you, my mommy.

My Sister . . .

You were my first and best friend. You taught me how to be a sister and a friend, which we will always be for the rest of our lives. . . . Love you, my sister.

Brad Cafarelli . . .

Couldn't do it without you and your laughter. Your wit, wisdom, and expertise are essential to everything we accomplish together, but they are nothing compared to your love and friendship.

Rick Yorn . . .

Thank you for being my partner for twenty-two years. Your guidance has always taken me to where I'm meant to be. I can't wait to see all the places we still have to go. . . . Love you beyond.

Sandra Bark . . .

You did it!! You wrestled this beast of a topic to the ground. . . . You pulled it apart, broke it down, and identified all its parts. You found its backbone and its heart and all its moving parts and then you put it all back together again. You are a genius. That mind of yours continues blow me away. I love our process, and am so thankful for our partnership. . . . I've learned so very much from you, dear friend.

All of the experts who generously offered their time . . .

Sandra and I learned so much on this journey, and all of that knowledge was gathered through conversations, by asking questions, by reading books and articles and then asking more questions. Thank you to all of the people who helped us satisfy our curiosity and usher us through to yet bigger questions. We have tried to list all of them here!

The National Institutes of Health was our university. Dr. Kayvon Modjarrad first talked to us about cells and senescence, and invited us to the National Institutes of Health, where we met and talked with Dr. Francis Collins, director of the NIH, Dr. Janine Clayton, director of the Offices of Research of Women's Health, Dr. Richard Hodes, director of the National Institute of Aging, Dr. Felipe Sierra, director of the NIA's Division of Aging Biology, Dr. Luigi Ferrucci, director of the Baltimore Longitudinal Study of Aging, along with Dr. Susan Resnick, Dr. John Mascola, Dr. John Gallin, Dr. Kong Chen, Dr. Leighton Chan, Dr. Diane Damiano, Dr. Andrew Singleton— all made possible with the support of John Burklow and Renate Myles.

We stretched our minds and our imaginations at the Buck Institute, at the invitation of Dr. Judith Campisi, and learned about the latest in aging science research from Dr. Brian Kennedy, director of the Buck Institute, and Dr. Julie Andersen, Dr. Simon Melov, Dr. Patricia Spillman, Dr. Gordon Lithgow, Dr. Henri Jasper, Dr. Pankaj Kapahi, with the support of Kris Rebillot.

At UCSF, Dr. Elissa Epel talked with us about stress and cellular aging. At UCLA, Dr. Eric Esrailian, Dr. Lin Chang, Nancee Jaffe, M.S., R.D., all talked with us about health and aging. Dr. Gail Greendale sat down with us to talk about the menopause transition. Dr. Dale Bredesen shared with us his research on Alzheimer's disease. The staff at the Iris Cantor-UCLA Women's Health Center invited us to lectures and luncheons.

So many people from so many institutions guided us along our path. Dr. Vincent Marchesi warned us about the dangers of oxygen and introduced us to proteins. Dr. Caleb Finch at USC gave us information and encouragement. We talked with fertility expert Dr. Diana Chavkin about our reproductive systems, with neurologist Dr. Shennan Weiss about our brains, with Dr. Seth Uretsky about healthy aging and our hearts, and with Dr. Sam Klein, director of the Center for Human Nutrition at Washington University about nutrition and aging.

To all of the people who gave their energy, time, and knowledge to this endeavor...

Thank you to everyone at HarperWave who allowed this book to exist and make its way into your hands, including Karen Rinaldi, Brian Perrin, Leslie Cohen, Lydia Weaver, Leah Carlson-Stanisic, and Kate Lyons.

A million thanks to Paul Kepple for once again creating such a gorgeous book design and to Harriet Russell for contributing your beautiful illustrations. Thank you to Jeff Dunas for your magical, visual storytelling. I can't wait for the next one.

A round of applause to Carlotta Duncan, PhD, who worked on nearly every stage of this manuscript. Thank you for your enthusiasm for this subject, support for this project, and unwavering attention to detail!! And three cheers for Joanna Parson at Letter Perfect Transcription who listened closely through hours and hours and hours of tape.

Our endless appreciation to Laurie David for allowing us to share the beautiful poem that closes this book.

And finally, a big shout-out of appreciation to all the Mommies, Aunties, Besties, Sissies, and Nieces out there: you are the foundation upon which all the strongest bonds are built. I don't know where I would be without these women in my life. I look forward to the journeys ahead of us all. . .

NOTES

INTRODUCTION

6   life expectancy is longer: Max Roser, "Life Expectancy," OurWorldInData.org (2015), http://ourworldindata.org/data/population-growth-vital-statistics/life-expectancy/, accessed December 3, 2015.

CHAPTER 1

15   sexually mature enough to reproduce: Roger B. McDonald, *Biology of Aging* (New York: Garland Science, 2013).

16   made of lead: Diane Mapes, "Suffering for beauty has ancient roots," NBC News, http://www.nbcnews.com/id/22546056/ns/health/t/suffering-beauty-has-ancient-roots, accessed November 18, 2015.

17   the ammonia in urine: Mohi Kumar, "From Gunpowder to Teeth Whitener: The Science Behind Historic Uses of Urine," Smithsonian.com (August 2013), http://www.smithsonianmag.com/science-nature/from-gunpowder-to-teeth-whitener-the-science-behind-historic-uses-of-urine-442390, accessed December 3, 2015.

16   Ended up in Florida: http://www.fountainofyouthflorida.com/, accessed November 18, 2015.

17   make a couple of facial incisions: "About face," *New York* magazine, http://nymag.com/news/features/48948/index2.html, accessed November 18, 2015.

17   first textbook on the subject: Paul Howard, MD, "The history of face lift," http://thehowardlift.com/facelifthistory.html, accessed November 18, 2015.

16   Botox: NBC News, http://www.nbcnews.com/id/21369061/ns/health-skin_and_beauty/t/frozen-time-botox-over-years, accessed November 18, 2015.

17   Bee venom: Madeline Vann, "7 Bizarre Anti-Aging Beauty Treatments," http://www.everydayhealth.com/skin-and-beauty-pictures/bizarre-anti-aging-beauty-treatments.aspx#01, accessed November 18, 2015.

17   placenta face masks: Ibid.

17   quit smiling for forty years in an attempt to avoid getting wrinkles: Kate Sullivan, "Woman Claims She Hasn't Smiled for 40 Years to Avoid Wrinkles," Allure magazine, http://www.

allure.com/beauty-trends/blogs/daily-beauty-reporter/2015/02/woman-wont-smile-beauty.html, accessed November 18, 2015.

18    $30 billion a year on cosmetics: Dodai Stewart, "Americans Spend Billions on Beauty Products But Are Not Very Happy," http://jezebel.com/5931654/americans-spend-billions-on-beauty-products-and-are-still-pretty-unhappy/, accessed November 18, 2015.

CHAPTER 2

25   children died regularly from fevers: "The History of Antibiotics," American Academy of Pediatrics, https://www.healthychildren.org/English/health-issues/conditions/treatments/Pages/The-History-of-Antibiotics.aspx, accessed December 3, 2015.

25   life-threatening infection: "Antibiotics 1928-2000," Australian Broadcasting Organization, http://www.abc.net.au/science/slab/antibiotics/history.htm, accessed December 3, 2015.

26   For millennia, smallpox: http://www.academia.edu/4573449/History_of_Smallpox _%C4%80bele_in_Persia. http://www.ncbi.nlm.nih.gov/pmc/articles/PMC1200696/.

26   Smallpox is even: http://www.ncbi.nlm.nih.gov/pmc/articles/PMC1200696/, accessed December 16, 2014.

26   400,000 people a year: http://www.historytoday.com/elizabeth-fenn/great-smallpox-epidemic, accessed January 21, 2015.

26   no cure: http://www.ncbi.nlm.nih.gov/pmc/articles/PMC1200696/.

26   a process called inoculation: http://www.historyofvaccines.org/content/timelines/smallpox, accessed December 16, 2014.

26   Variolation: Ibid.

26   Lady Mary Montagu: ibid.http://www.ncbi.nlm.nih.gov/pmc/articles/PMC1200696/.27 bacteria cause illnesses like cholera: "Bacteria: Life History and Ecology," http://www.ucmp.berkeley.edu/bacteria/bacterialh.html, accessed November 5, 2014.

27   puerperal fever: "Puerperal fever," *Encyclopedia Britannica*, http://www.britannica.com/EBchecked/topic/534198/Ignaz-Philipp-Semmelweis, accessed December 3, 2015.

28   maternal and infant mortality rates: Imre Zoltán, M.D., "Ignaz Philipp Semmelweis," *Encyclopedia Britannica*, http://www.britannica.com/EBchecked/topic/534198/Ignaz-Philipp-Semmelweis, accessed November 18, 2015.

29   the idea of antisepsis: Ibid.

29   children, young adults, and women: "Global health and aging," National Institutes of Aging, https://www.nia.nih.gov/research/publication/global-health-and-aging/living-longer, accessed November 18, 2015.

29   number one killer: "Cardiovascular diseases," WHO (January 2015), http://www.who.int/mediacentre/factsheets/fs317/en/, accessed December 3, 2015.

30   When Life Got Longer: Department of Health and Human Services, National Center for Health Statistics, Centers for Disease Control and Prevention; National Vital Statistics Reports, www.dhhs.gov; www.cdc.gov.

30   Framingham Heart Study: "Framingham Heart Study," NHLBI/ Boston University, https://www.framinghamheartstudy.org, accessed December 3, 2015.

30 gave doctors a way: Nirav J. Mehta and Ijaz A. Khan, "Cardiology's 10 Greatest Discoveries of the 20th Century," *Texas Heart Institute Journal* 29, no. 3 (2002): 164-71, http://www.ncbi.nlm.nih.gov/pmc/articles/PMC124754/.

30 first major study that included women: Ibid.

30 mainly aimed at easing the symptoms: Roger B. McDonald, *Biology of Aging* (New York: Garland Science, 2013).

31 Doctors better understood: "History of medicine," *Encyclopedia Britannica,* http://www.britannica.com/EBchecked/topic/372460/history-of-medicine.

31 By the 1960s, heart transplant: Ibid.

31 first human heart was transplanted: The History Channel, http://www.history.com/this-day-in-history/first-human-heart-transplant.31

31 deaths dropped by half: John P. Bunker, "The role of medical care in contributing to health improvements within societies," *International Journal of Epidemiology* 30, no.6 (2001): 1260-3, http://ije.oxfordjournals.org/content/30/6/1260.long.

31 human life expectancy got longer: Ibid.

32 nearly 20 percent of American adults still smoke: "Current Cigarette Smoking Among Adults in the United States," Center for Disease Control and Prevention, http://www.cdc.gov/tobacco/data_statistics/fact_sheets/adult_data/cig_smoking/, accessed November 22, 2015.

32 Nearly 70 percent of adults are obese: "Fast Facts on the State of Obesity in America," The State of Obesity, http://stateofobesity.org/fastfacts, accessed November 22, 2015.

32 Less than 40 percent eat the recommended: Patricia M. Guenther et al., "Most Americans eat much less than recommended amounts of fruits and vegetables," *Journal of the Academy of Nutrition and Diabetics* 106, no. 9 (September 2006): 1371-79, http://www.ncbi.nlm.nih.gov/pubmed/16963342.

33 there are now more people over the age of sixty-five: "NIH Global Health and Aging," October 2011, National Institutes of Health, http://www.nia.nih.gov/sites/default/files/global_health_and_aging.pdf.

34 Sixty-six percent of caregivers are female: "Women and Caregiving: Facts and Figures," Family Caregiver Alliance, https://www.caregiver.org/women-and-caregiving-facts-and-figures, accessed November 22, 2015.

35 valued from $148 billion to $188 billion: Ibid.

35 long-term caregivers are likely to suffer: Elissa S. Epel, "Accelerated telomere shortening in response to life stress," *Proceedings of the National Academy of Sciences* 101, no. 49 (December 7, 2004): 17312–17315, http://www.pnas.org/content/101/49/17312.long.

35 more likely to retire: "Women and Caregiving: Facts and Figures," Family Caregiver Alliance, https://www.caregiver.org/women-and-caregiving-facts-and-figures, accessed November 22, 2015.

CHAPTER 3

36 the National Institute on Aging: "NIA timeline," National Institute on Aging, http://www.nia.nih.gov/about/nia-timeline.

36 funder of Alzheimer's research: National Institute on Aging, https://www.nia.nih.gov/, accessed November 22, 2015.

36 one-billion- dollar budget for research grants: National Institute on Aging, https://www.nia.nih.gov/about/budget/2015/fiscal-year-2016-budget, accessed November 22, 2015.

36 $25 million to study aging: "Buck Institute Awarded $25 Million to Establish New Scientific Discipline of Geroscience," Buck Institute for Research on Aging (September 2007), http://www.buckinstitute.org/buck-news/buck-institute-awarded-25-million-establish-new-scientific-discipline-geroscience, accessed December 4, 2015.

37 collaborative paper called "Geroscience": Brian K. Kennedy et al., "Geroscience: Linking Aging to Chronic Disease," Cell 159, No. 4 (November 6, 2014): 709-13, http://www.sciencedirect.com/science/article/pii/S009286741401366X.

37 Geras—which translates: H. G. Liddell and R. Scott, "A Greek-English Lexicon" (Oxford: Clarendon Press, 1940), http://perseus.uchicago.edu/Reference/LSJ.html.

38 aging is the biggest single risk factor: Kennedy, "Geroscience."

38 Buck Institute's Interdisciplinary Research Consortium: Ibid.

39 three kinds of primary cell layers: "Embryonic stem cell," ScienceDaily, http://www.sciencedaily.com/terms/embryonic_stem_cell.htm, accessed November 22, 2015.

39 adults have stem cells: "What are adult stem cells?," National Institutes of Health, http://stemcells.nih.gov/info/basics/pages/basics4.aspx, accessed November 22, 2015.

39 potential to morph: Ibid.

39 Some adult stem cells can also be activated: "Exercise triggers stem cells in muscle," ScienceDaily (February 6, 2012), http://www.sciencedaily.com/releases/2012/02/120206143944.htm.

39 The 2012 Nobel Prize: "The Nobel Prize in Physiology or Medicine 2012." Nobel Media AB 2014. http://www.nobelprize.org/nobel_prizes/medicine/laureates/2012/, accessed November 22, 2015.

39 can be derived from specific people: "What are adult stem cells?," National Institutes of Health, http://stemcells.nih.gov/info/basics/pages/basics4.aspx, accessed November 22, 2015.

40 aging at the cellular level: Kennedy, "Geroscience."

40 longitudinal studies: "Nursing Resources: Types of Studies," University of Wisconsin Madison – Health Sciences, http://researchguides.ebling.library.wisc.edu/c.php?g=293229&p=1953448.

40 cross-section of people: Ibid.

41 vitamin D deficiency: Consuelo H. Wilkins et al., "Vitamin D deficiency is associated with low mood and worse cognitive performance in older adults," American Journal of Geriatric Psychiatry 14, No. 12 (December 2006):1032-40, http://www.ncbi.nlm.nih.gov/pubmed/17138809.

41 interventional study: Ibid.

41 any exercise had great benefits: R. Ruscheweyh et al., "Physical activity and memory functions: an interventional study," Neurobiology of Aging 32, No. 7, (July 2011): 1304-19, http://www.ncbi.nlm.nih.gov/pubmed/19716631.

41 from sponges and worms: McDonald, Aging.

41   many genes and biological mechanisms: "Biology of Aging," National Institute on Aging (November 2011), https://www.nia.nih.gov/health/publication/biology-aging/.

42   Naked mole rats: Ibid.

42   By comparing similar animals: "Biology of Aging," National Institute on Aging.

CHAPTER 4

44   A baby girl born in the United States in 2010: "Life Expectancy at Birth (in years), by Gender," The Henry J. Kaiser Family Foundation, http://kff.org/other/state-indicator/life-expectancy-by-gender/, accessed November 27, 2015.

44   the world over, women's life expectancies: "World Health Statistics 2014," WHO, http://www.who.int/mediacentre/news/releases/2014/world-health-statistics-2014/en/.

44   women use more healthcare services: Klea D. Bertakis et al., "Gender differences in the utilization of health care services," *The Journal of Family Practice* 49, no. 2 (February 2000):147-52, http://www.ncbi.nlm.nih.gov/pubmed/10718692.

44   nearly 70 percent of Americans: Darren McCollester, "Study shows 70 percent of Americans take prescription drugs," CBS News, http://www.cbsnews.com/news/study-shows-70-percent-of-americans-take-prescription-drugs/, accessed November 27, 2015.

45   Women are more likely to visit the doctor: "United States, 2006, with chartbook on trends in the health of Americans," National Center for Health Statistics, Centers for Disease Control (CDC), (2006) www.cdc.gov/nchs/data/hus/hus06.pdf .

45   healthy choices that... women make: "Mars vs. Venus: The gender gap in health," Harvard Health Publications, http://www.health.harvard.edu/newsletter_article/mars-vs-venus-the-gender-gap-in-health, accessed December 5, 2015.

45   unintentional injury is number three: "Leading Causes of Death in Males United States, 2013," CDC, http://www.cdc.gov/men/lcod/2013/index.htm, accessed November 27, 2015.

45   number six for women: "Leading Causes of Death in Females United States, 2013," CDC, http://www.cdc.gov/Women/lcod/2013/index.htm, accessed November 27, 2015.

46   chimpanzees: "How long do chimpanzees live?," Chimpanzee Sanctuary Northwest, http://www.chimpsanctuarynw.org/blog/2013/03/how-long-do-chimpanzees-live/.

46   maximum human life: McDonald, Roger, *The Biology of Aging.*

46   Jeanne Calment: Jonathan Silvertown, *The Long and the Short of It: The Science of Life Span and Aging* (Chicago, IL: The University of Chicago Press, 2013).

46   Cholesterol plaques can settle: "A to Z Guide: Sex and Gender Influences on Health," NIH Office of Research on Women's Health (ORWH), http://orwh.od.nih.gov/resources/sexgenderhealth/index.asp#c, accessed November 27, 2015.

47   more women than men die of cardiovascular disease: "Sex Chromosomes," John W.Kimball, Kimball's Biology Pages, http://users.rcn.com/jkimball.ma.ultranet/BiologyPages/S/SexChromosomes.html,accessed December 1, 2015.

47   defibrillators—which have primarily been tested on men: Leslie Mann, "Women more likely to have complications from implantable cardiac defibrillators, study says," (January 2013), http://articles.chicagotribune.com/2013-01-09/health/ct-x-0109-women-defibrillators-20130109_1_defibrillators-heart-disease-sudden-cardiac-death.

47   more likely than men to develop depression: "Common Mental Health Issues Among Women," dualdiagnosis.org, http://www.dualdiagnosis.org/mental-health-and-addiction/common-issues-women/; "Posttraumatic Stress Disorder (PTSD)," Anxiety and Depression Association of America (ADAA), http://www.adaa.org/understanding-anxiety/posttraumatic-stress-disorder-ptsd; "Women," ADAA, http://www.adaa.org/living-with-anxiety/women.

47   twice the risk of having a stroke: "Depression Increases Risk of Stroke in Women," Go Red for Women, American Heart Association, https://www.goredforwomen.org/about-heart-disease/heart_disease_research-subcategory/depression-increases-risk-of-stroke-in-women/, accessed December 1, 2015.

48   forty-six chromosomes: "Sex Chromosomes," John W. Kimball, Kimball's Biology Pages, http://users.rcn.com/jkimball.ma.ultranet/BiologyPages/S/SexChromosomes.html, accessed December 1, 2015.

49   sex chromosome carried by the sperm: "Sex chromosome," *Encyclopedia Britannica*, http://www.britannica.com/EBchecked/topic/536952/sex-chromosome, accessed November 27, 2015.

49   if one of our Xs contains a faulty gene: David Robson, "Why do women live longer than men?," BBC News (October 2015) http://www.bbc.com/future/story/20151001-why-women-live-longer-than-men.

49   red-green color blindness: "Studies Expand Understanding of X Chromosome," National Human Genome Research Institute, https://www.genome.gov/13514331, accessed November 27, 2015.

50   "hysteria" is actually derived: "Hysteria,"Oxford Dictionaries, http://www.oxforddictionaries.com/us/definition/american_english/hysteria, accessed November 27, 2015.

50   Married women were told: Rachel P Maines, *The Technology of Orgasm: "Hysteria," the Vibrator, and Women's Sexual Satisfaction* (Baltimore, MD: Johns Hopkins University Press, 1999).

50   stimulate women to orgasm: Ibid.

50   nearly three-quarters of the female population: Ibid.

50   "hysterical neurosis" was finally dropped: Cecilia Tasca Women, "Hysteria in the History of Mental Health," *Clinical Practice & Epidemiology in Mental Health* 8 (October 2012) 110–119, http://www.ncbi.nlm.nih.gov/pmc/articles/PMC3480686/.

50   bikini medicine: "Women's Health: More Than 'Bikini Medicine,'" National Public Radio, http://www.npr.org/2013/03/25/175267713/womens-health-more-than-bikini-medicine, accessed November 27, 2015.

51   30 percent of ob-gyn specialists were women: Carol S. Weisman, "Changing Definitions of Women's Health: Implications for Health Care and Policy," *Maternal and Child Health Journal* 1, No. 3 (September 1997): 179-189, http://link.springer.com/article/10.1023/A%3A1026225513674.

52   Margaret Sanger opens: "Women's Health Matters," UCSF Women's Health (Winter 2001) http://www.whrc.ucsf.edu/whrc/healthed/nwlswinter2001.pdf.

53   At the urging of the FDA: "FDA Drug Safety Communication: Risk of next-morning impairment after use of insomnia drugs; FDA requires lower recommended doses for certain drugs containing zolpidem," U.S. Food and Drug Administration, FDA http://www.fda.gov/Drugs/DrugSafety/ucm334033.htm, accessed November 27, 2015.

53   NIH calls for female cells: Janine A. Clayton and Francis S. Collins, "Policy: NIH to balance sex in cell and animal studies," *Nature* 509 (May 2014): 282–283, http://www.nature.com/news/policy-nih-to-balance-sex-in-cell-and-animal-studies-1.15195.

54   drugs don't affect men and women: "Chapter 3: Physiological Effects of Alcohol, Drugs, and Tobacco on Women," *Substance Abuse Treatment: Addressing the Specific Needs of Women* (Rockville, MD: Substance Abuse and Mental Health Services Administration, US: 2009), http://www.ncbi.nlm.nih.gov/books/NBK83244/.

54   A female liver metabolizes: Heather P. Whitley and Wesley Lindsey, "Sex-Based Differences in Drug Activity," *American Family Physician* (AFP) 1, No. 11 (December 2009):1254-1258 http://www.aafp.org/afp/2009/1201/p1254.html.

54   BODY COMPOSITION: Ibid.

55   FEMALE HORMONES: Ibid.

55   30 percent higher sensitivity: Ibid.

55   males and females do not respond in similar ways to opioids: A to Z Guide, NIH ORWH.

55   rise in the numbers of women dying: Nora Volkow, "Prescription Painkillers Are Claiming More Women's Lives," NIH National Institute on Drug Abuse (NIDA), https://www.drugabuse.gov/about-nida/noras-blog/2013/07/prescription-painkillers-are-claiming-more-womens-lives, accessed November 27, 2015.

55   women between the ages of forty-five to fifty-four: "Telebriefing on Deaths from Prescription Painkiller Overdoses Rise Sharply Among Women," (July 2013), http://www.cdc.gov/media/releases/2013/t0702-drug-overdose.html, accessed November 27, 2015.

55   sex isn't always taken into consideration: Marguerite Del Giudice, "Why It's Crucial to Get More Women Into Science," *National Geographic* (November 2014), http://news.nationalgeographic.com/news/2014/11/141107-gender-studies-women-scientific-research-feminist/, accessed November 27, 2015.

56   Pregnancy is a concern: Joanne Cavanaugh Simpson, "Pregnant pause," *Johns Hopkins* magazine, http://pages.jh.edu/~jhumag/0901web/pregnant.html, accessed November 27, 2015.

56   half as much flu vaccine: A to Z Guide, NIH ORWH.

57   thirty million women between the ages of thirty-five and fifty: Lindsay M. Howden and Julie A. Meyer, "Age and sex contribution: 2010," United States Census Bureau (May 2011), http://www.census.gov/prod/cen2010/briefs/c2010br-03.pdf.

CHAPTER 5

63   some research ties longevity to the genes we inherit: Nir Barzilai, "The Longevity Genes Project," Albert Einstein College of Medicine, https://www.einstein.yu.edu/centers/aging/longevity-genes-project/, accessed November 27, 2015.

63   our choices, our environment: "Lifestyle affects life expectancy more than genetics, Swedish study finds, ScienceDaily (February 2011) http://www.sciencedaily.com/releases/2011/02/110207112539.htm.

64   your genotype: "Genotype," Genetics Home Reference, U.S. NLM http://ghr.nlm.nih.gov/glossary=genotype, accessed December 6, 2015.

65   abuse, neglect, or chaos: Alexandra D. Crosswell et al., "Childhood Adversity and In-flammation in Breast Cancer Survivors," *Psychosomatic Medicine* 76, No. 3 (April 2014): 208–214. http://www.ncbi.nlm.nih.gov/pmc/articles/PMC4357419/.

65   alter the way your genes are expressed: Alice G. Walton, "How Health and Lifestyle Choices Can Change Your Genetic Make-Up," *The Atlantic Monthly* (November 2011), http://www.theatlantic.com/health/archive/2011/11/how-health-and-lifestyle-choices-can-change-your-genetic-make-up/247808/.

65   flipped on or off by experience: "The Epigenome learns from its experiences," University of Utah Health Sciences, http://learn.genetics.utah.edu/content/epigenetics/epi_learns/, accessed December 6, 2015.

65   passed down to the next generation: "Epigenetics and Inheritance," University of Utah Health Sciences, http://learn.genetics.utah.edu/content/epigenetics/inheritance/, accessed December 6, 2015.

65   epigenetics: "Epigenetics," University of Utah Health Sciences, http://learn.genetics.utah.edu/content/epigenetics/, accessed December 6, 2015.

66   phenotype: "phenotype / phenotypes," SciTable, Nature Education, http://www.nature.com/scitable/definition/phenotype-phenotypes-35, accessed December 6, 2015.

66   identical twin: Ibid.

67   onset of aging happens: MacDonald, *Biology of Aging.*

67   a little bit less oxygen: "Biology of Aging," National Institute on Aging.

67   cognitive skills also begin to slip: Alice Park, "Our Brains Begin to Slow Down at Age 24," *Time* magazine (April 2014), http://time.com/63500/brain-aging-at-24/, accessed November 27, 2015.

67   mass in your muscles: "Sarcopenia with aging," WebMD, http://www.webmd.com/healthy-aging/sarcopenia-with-aging, accessed December 6, 2015.

68   filter blood less efficiently: Richard W. Besdine, "Changes in the Body With Aging," Merck Manuals, http://www.merckmanuals.com/home/older-people-s-health-issues/the-aging-body/changes-in-the-body-with-aging, accessed December 6, 2015.

68   peak bone mass: "Healthy Bones at Every Age," American Academy of Orthopaedic Surgeons, http://orthoinfo.aaos.org/topic.cfm?topic=a00127, accessed December 6, 2015.

68   lens of the eye thickens: Gretchyn Bailey, "Presbyopia," http://www.allaboutvision.com/conditions/presbyopia.htm, accessed November 27, 2015.

68   decreased melanin production: "Everyday mysteries: Why does hair turn grey?," Library of Congress, https://www.loc.gov/rr/scitech/mysteries/grayhair.html, accessed November 27, 2015.

68   decreased collagen production: Angelica Carrillo Leal, "Why Does Your Skin Age?," *Dartmouth Undergraduate Journal of Science* (January 2013), http://dujs.dartmouth.edu/news/why-does-your-skin-age#.VmS45cpG_2g, accessed December 6, 2015.

68   liver can no longer process alcohol: Andrea Petersen, "Drinking After 40: Why Hangovers Hit Harder," *Wall Street Journal* (WSJ) (November 2013), http://www.wsj.com/articles/SB10001424052702304439804579205913000870266, accessed December 6, 2015.

68   senescence: "Senescent," Dictionary.com, http://dictionary.reference.com/browse/senescent, accessed December 6, 2015.

68   senescent cells: Ibid.

68   less muscle, bone, and fat beneath your skin: Beth Howard, "What to Expect in Your 50s," *AARP The Magazine* (October 2012), http://www.aarp.org/health/healthy-living/info-09-2012/what-to-expect-in-your-50s.2.html.

69   In your sixties: Howard, "What to Expect in Your 60s," *AARP*, http://www.aarp.org/health/healthy-living/info-09-2012/what-to-expect-in-your-60s.2.html.

69   Alzheimer's disease: Alzheimer's Association, "2014 Alzheimer's Disease," *Alzheimer's & Dementia* 10, No. 2 (2014) Facts and Figures," http://www.alz.org/downloads/facts_figures_2014.pdf.

69   In your seventies: Beth Howard, "What to Expect in Your 70s and beyond," AARP, http://www.aarp.org/health/healthy-living/info-09-2012/what-to-expect-in-your-70s-and-beyond.html.

69   leading cause of death for people aged seventy-five to eighty-four: Ibid.

69   In your eighties: "Healthy aging into your 80s and beyond," *Consumer Reports* (May 2014), http://www.consumerreports.org/cro/magazine/2014/06/healthy-aging-into-your-80s-and-beyond/index.htm, accessed December 6, 2015.

70   happiest people: Mindy Greenstein and Jimmie Holland, *Lighter as We Go: Virtues, Character Strengths, and Aging* (New York: Oxford University Press, 2015), http://www.amazon.com/dp/0199360952/.

70   stress and worry start to decrease: Frank Newport and Brett Pelham, "Don't Worry, Be 80: Worry and Stress Decline With Age," Gallup Poll (December 2009), http://www.gallup.com/poll/124655/dont-worry-be-80-worry-stress-decline-age.aspx, accessed December 6, 2015.

71   people who live to be one hundred: Kaori Kato et al., "Positive attitude towards life and emotional expression as personality phenotypes for centenarians," *Aging* (Albany NY) 4, No. 5 (May 2012): 359–367, http://www.ncbi.nlm.nih.gov/pmc/articles/PMC3384436/.

71   eat well, work out: Becca R. Levy and Lindsey M. Myers, "Preventive health behaviors influenced by self-perceptions of aging," *Preventive Medicine* 39, No. 3 (September 2004): 625–629, http://www.ncbi.nlm.nih.gov/pubmed/15313104.

71   Nun Study: Deborah D. Danner et al., "Positive Emotions in Early Life and Longevity: Findings from the Nun Study," *Journal of Personality and Social Psychology* 80, No. 5, (May 2001): 804-813, https://www.apa.org/pubs/journals/releases/psp805804.pdf.

71   live approximately 7.5 years longer: Becca R. Levy et al., "Longevity increased by positive self-perceptions of aging," *Journal of Personality and Social Psychology* 83, No. 2 (August 2002): 261-270, http://www.ncbi.nlm.nih.gov/pubmed/12150226.

72   embrace the idea of aging age: Baltimore Longitudinal Study of Aging, NIA, https://www.blsa.nih.gov/.

72   your DNA: Kato, *Aging*.

72   the kind of gains usually seen from exercise: Becca R. Levy, "Mind Matters: Cognitive and Physical Effects of Aging Self-Stereotypes," *The Journals of Gerontology*: Series B 58, No. 4: 203-211, http://psychsocgerontology.oxfordjournals.org/content/58/4/P203.full.

CHAPTER 6

74 forgot the sunscreen: "What causes our skin to age?," American Academy of Dermatology, https://www.aad.org/dermatology-a-to-z/health-and-beauty/every-stage-of-life/adult-skin/what-causes-aging-skin, accessed November 28, 2015.

75 (UV) radiation: "Wrinkles," Mayo Clinic, http://www.mayoclinic.org/diseases-conditions/wrinkles/basics/causes/con-20029887, accessed November 28, 2015.

75 Exposure to UV light: "Age spots (liver spots)," Mayo Clinic, http://www.mayoclinic.org/diseases-conditions/age-spots/basics/definition/con-20030473, accessed November 28, 2015.

75 the make energy: "What is a cell?," Genetics Home Reference, U.S. NLM, http://ghr.nlm.nih.gov/handbook/basics/cell, accessed December 6, 2015.

76 lead to chronic diseases: Denham Harman, "Free radical theory of aging," *Mutation Research/DNAging* 275, No. 3–6 (September 1992): 257-266, http://www.ncbi.nlm.nih.gov/pubmed/1383768.

76 CELL MEMBRANE: Alexandra Villa-Forte, "Cells," Merck Manuals, http://www.merckmanuals.com/home/fundamentals/the_human_body/cells.html, accessed November 28, 2015.

76 LYSOSOMES: Ibid.

76 ORGANELLES: Ibid.

76 MITOCHONDRIA, the powerhouses: Ibid.

78 energy that fuels your life: Faculty of Education, "Mitochondria – cell powerhouses," University of Waikato, New Zealand, http://sciencelearn.org.nz/Contexts/Digestion-Chemistry/Looking-Closer/Mitochondria-cell-powerhouses, accessed November 28, 2015.

78 free radicals: Ibid.

78 misfolded proteins: Northwestern University, "Misfolded Proteins: The Fundamental Problem Is Aging," ScienceDaily (August 2009) http://www.sciencedaily.com/releases/2009/08/090824151251.htm, accessed December 6, 2015.

79 cells that do not divide correctly: McDonald, *Biology of Aging*.

79 antioxidants . . . may be harmful: Ibid.

80 mitochondria in your cells use 95 percent: Roland N. Pittman, *Regulation of Tissue Oxygenation* (San Rafael, CA: Morgan & Claypool Life Sciences, 2011), http://www.ncbi.nlm.nih.gov/books/NBK54110/.

80 Without this neat biological trick: McDonald, *Biology of Aging*.

80 producing more mitochondria: "Mitochondria," *Nature Education* (2014), http://www.nature.com/scitable/topicpage/mitochondria-14053590, accessed November 28, 2015.

80 molecules without electrons: McDonald, *Biology of Aging*.

81 Vitamins C and E can behave: "Antioxidants: Beyond the Hype," Harvard T.H. Chan School of Public Health, http://www.hsph.harvard.edu/nutritionsource/antioxidants/, accessed December 6, 2015.

81 researchers compared the mitochondrial function: E.J. Brierley et al., "Mitochondrial function in muscle of elderly athletes," *Annals of Neurology* 41, No. 1(January 1997):114-6, http://www.researchgate.net/publication/14202422_Mitochondrial_function_in_muscle_of_elderly_athletes.

81   brain cells require a lot of energy: Nikhil Swaminathan, "Why Does the Brain Need So Much Power?," *Scientific American* (April 2008) http://www.scientificamerican.com/article/why-does-the-brain-need-s/.

81   brain demands 20 percent: "Neuroscience For Kids: The Blood Supply of the Brain," University of Washington, https://faculty.washington.edu/chudler/vessel.html, accessed December 6, 2015.

81   part of what causes Alzheimer's disease: Paula I. Moreira et al., "Mitochondrial dysfunction is a trigger of Alzheimer's disease pathophysiology," *Biochimica et Biophysica Acta (BBA) - Molecular Basis of Disease* 1802, No. 1 (January 2010): 2–10 http://www.sciencedirect.com/science/article/pii/S09254439

82   they used to be independent cells: "Mitochondria," Scitable, *Nature Education,* http://www.nature.com/scitable/topicpage/mitochondria-14053590.

82   *All multicelled organisms*: "Eukaryotic Cells," Scitable, *Nature Education*, http://www.nature.com/scitable/topicpage/eukaryotic-cells-14023963, accessed December 6, 2015.

82   20-25,000 total genes: "A Brief Guide to Genomics," National Human Genome Research Institute (NHGRI), http://www.genome.gov/18016863, accessed November 28, 2015.

82   only thirty-seven genes live in our mitochondria: Heidi Chai and Joanna Craig, "mtDNA and mitochondrial diseases," *Nature Education* 1, No. 1 (2008):217, http://www.nature.com/scitable/topicpage/mtdna-and-mitochondrial-diseases-903.

83   the strand is six feet long: "Chromosomes," National Human Genome Research Institute (NHGRI), http://www.genome.gov/26524120, accessed November 28, 2015.

83   stomach, where our cells are replaced every five days: Nicholas Wade, "Your Body Is Younger Than You Think," *New York Times* (August 2005), http://www.nytimes.com/2005/08/02/science/your-body-is-younger-than-you-think.html?_r=0.

83   mutation: "What is mutation?," University of Utah Health Sciences, http://learn.genetics.utah.edu/content/variation/mutation/, accessed December 6, 2015.

83   telomeres: "Are Telomeres the Key to Aging and Cancer?," University of Utah Health Sciences, http://learn.genetics.utah.edu/content/chromosomes/telomeres/, accessed December 6, 2015.

84   telomerase, an enzyme: "The Nobel Prize in Physiology or Medicine 2009: Elizabeth H. Blackburn, Carol W. Greider, Jack W. Szostak," Nobel Prize.org, (October 2009), http://www.nobelprize.org/nobel_prizes/medicine/laureates/2009/press.html.

84   length of your telomeres: Ibid.

84   telomeres are too short: "Are Telomeres the Key to Aging and Cancer?," University of Utah Health Sciences, http://learn.genetics.utah.edu/content/chromosomes/telomeres/, accessed December 6, 2015.

84   women who experienced chronic stress: Elissa S. Epel et al., "Accelerated telomere shortening in response to life stress," PNAS 101, no. 49 (December 2004): 17312–17315, http://www.pnas.org/content/101/49/17312.full.pdf+html.

85   telomeres can also be shortened by poor nutrition: Jue Lin et al., "Telomeres and lifestyle factors: Roles in cellular aging," Mutation Research/Fundamental and Molecular Mechanisms of Mutagenesis 730 (August 2011): 85– 89, http://www.pnas.org/content/101/49/17312.full.pdf+html.

85   longer telomeres than men: Michael Gardner et al., "Gender and telomere length: systematic review and meta-analysis," *Experimental Gerontology* 51 (March 2014):15-27, http://www.ncbi.nlm.nih.gov/pubmed/24365661.

85   crucial protector against cancer: Francis Rodier and Judith Campisi, "Four faces of cellular senescence" *Journal of Cellular Biology* 192, No. 4 (February 2011): 547–556, http://www.ncbi.nlm.nih.gov/pmc/articles/PMC3044123/.

86   your immune system detects and clears senescent cells regularly: Lisa Hoenicke and Lars Zender, "Immune surveillance of senescent cells—biological significance in cancer and non-cancer pathologies," *Carcinogensis* (2012), http://carcin.oxfordjournals.org/content/early/2012/04/16/carcin.bgs124.full.

86   continue to accumulate as the years go by: Gregg Easterbrook, "What Happens When We All Live to 100?," *The Atlantic* (October 2014), http://www.theatlantic.com/magazine/archive/2014/10/what-happens-when-we-all-live-to-100/379338/.

86   killing senescent cells stops: Nicholas Wadenov, "Purging Cells in Mice Is Found to Combat Aging Ills," *New York Times* (November 2011) http://www.nytimes.com/2011/11/03/science/senescent-cells-hasten-aging-but-can-be-purged-mouse-study-suggests.html?_r=0.

86   "necrosis": "Necrosis", Medline Plus, U.S. NLM, https://www.nlm.nih.gov/medlineplus/ency/article/002266.htm, accessed December 6, 2015.

86   "Apoptosis": Bruce Alberts et al., *Molecular Biology of the Cell,* 4th Edition (New York: Garland Science, 2002), http://www.ncbi.nlm.nih.gov/books/NBK26873/.

86   a dead cell gives sustenance to other cells as it goes: Ibid.

86   fallen leaf that nourishes the tree: "Want To Improve Your Lawn? Don't Bag Those Leaves," NPR (October 2011) http://www.npr.org/2011/10/28/141761525/want-to-improve-your-lawn-dont-bag-those-leaves.

87   may contribute to the formation of tumors: Albert R Davalos et al., "Senescent cells as a source of inflammatory factors for tumor progression," *Cancer Metastasis Reviews* 29, No. 2 (June 2010):273-83.

87   ability to heal wounds: Campisi, *Journal of Cellular Biology.*

CHAPTER 7

91   ALL ABOUT YOUR HORMONES: "Endocrine Glands and Types of Hormones," The Hormone Network, Endocrine Society, http://www.hormone.org/hormones-and-health/the-endocrine-system/endocrine-glands-and-types-of-hormones, accessed November 28, 2015.

91   Greek *horman*: "hormone," *The American Heritage Science Dictionary* (Boston, MA: Houghton Mifflin Company, 2002), http://dictionary.reference.com/browse/hormone, accessed November 30, 2015.

91   As we grow and develop: "Aging changes in hormone production," MedlinePlus, U.S. National Library of Medicine (NLM), http://www.nlm.nih.gov/medlineplus/ency/article/004000.htm, accessed November 28, 2015.

91   endocrine system: "Endocrine Glands and Types of Hormones," Hormone Health Network, http://www.hormone.org/hormones-and-health/endocrine-glands-and-types-of-hormones, accessed December 7, 2015.

92   rises during childhood: "Growth hormone, athletic performance, and aging," *Harvard Men's Health Watch,* Harvard Medical School, http://www.health.harvard.edu/diseases-and-conditions/growth-hormone-athletic-performance-and-aging, accessed November 28, 2015.

92   produces follicle-stimulating hormone (FSH): "Pituitary gland," MedlinePlus, U.S. National Library of Medicine, https://www.nlm.nih.gov/medlineplus/ency/anatomyvideos/000099.htm, accessed November 28, 2015.

92   blood tests for FSH are one method: PonJola Coney, "Menopause," Medscape http://emedicine.medscape.com/article/264088-overview, accessed November 28, 2015.

92   more common among people over the age of sixty: "The Thyroid and You: Coping with a Common Condition," *NIH Medline Plus Magazine* 7, No. 1 (Spring 2012): 22-23, https://www.nlm.nih.gov/medlineplus/magazine/issues/spring12/articles/spring12pg22-23.html.

92   ADH levels rise as we age: John P. Kugler and Thomas Hustead, "Hyponatremia and Hypernatremia in the Elderly," *American Family Physician* 61, No. 12 (June 2000):3623-3630, http://www.aafp.org/afp/2000/0615/p3623.html, accessed December 5, 2015.

92   Elevated ADH levels also make elderly people: Kenneth L. Becker (Ed.), *Principles and Practice of Endocrinology and Metabolism* (Philadelphia, PA: Lippincott Williams & Wilkins, 2001), https://books.google.com/books?id=FVfzRvaucq8C.

92   regulation of menstrual cycles: "Menopause," University of Maryland Medical Health Center, https://umm.edu/health/medical/altmed/supplement/melatonin, accessed November 28, 2015.

93   during pregnancy: "Thyroid disease," Office on Women's Health, Thyroid and pregnancy, http://www.womenshealth.gov/publications/our-publications/fact-sheet/thyroid-disease.html, accessed December 5, 2015.

93   postpartum thyroiditis: thyroiditism, http://www.womenshealth.gov/publications/our-publications/fact-sheet/thyroid-disease.html.

93   older adults: Leslie M.C. Goldenberg, "Thyroid Disease in Late Life," Thyroid Foundation of Canada, http://www.thyroid.ca/e4g.php, accessed December 5, 2015.

93   SYMPTOMS: Ibid.

93   heart disease . . . osteoporosis: http://www.womenshealth.gov/publications/our-publications/fact-sheet/thyroid-disease.html.

93   effective medications: Ibid.

94   reproductive organs are a pretty pink: Natalie Angier, *Woman: An Intimate Geography* (New York: Houghton Mifflin Company, 1999), http://www.amazon.com/Woman-Intimate-Geography-Natalie-Angier/dp/0544228103.

94   when you were just a toddler: Ibid.

95   ovaries released estrogen and progesterone: Ibid.

95   doses of estrogen, progesterone, and testosterone: Gabrielle Lichterman, "Your Life in Hormones; *Marie Claire* magazine (April 2009), http://www.marieclaire.com/health-fitness/advice/a2853/female-hormones-cycle/.

96   in your thirties: Natalie Angier, *Woman*.

96   reserve of about three hundred thousand eggs: "Age and Fertility," American Society for Reproductive Medicine, https://www.asrm.org/BOOKLET_Age_And_Fertility/, accessed December 7, 2015.

96   aging is accompanied by a decline in fertility: "Infertility and Fertility: Overview," National Institute of Child Health and Human Development, http://www.nichd.nih.gov/health/topics/infertility/Pages/default.aspx, accessed November 28, 2015.

97   childbirth reduced breast cancer risk: Collaborative Group on Hormonal Factors in Breast Cancer, "Breast cancer and breastfeeding: collaborative reanalysis of individual data from

47 epidemiological studies in 30 countries, including 50302 women with breast cancer and 96973 women without the disease," *Lancet* 360, No. 9328 (July 2002):187-95, www.ncbi.nlm.nih.gov/pubmed/12133652.

97   decreased risk for ovarian cancer: Kim N. Danforth, "Breastfeeding and risk of ovarian cancer in two prospective cohorts," *Cancer Causes & Control* 18, No. 5, (June 2007): 517-23, http://www.ncbi.nlm.nih.gov/pubmed/17450440.

97   lactation stops ovulation: Alison M. Stuebe and E. Bimla Schwarz, "The risks and benefits of infant feeding practices for women and their children," *Journal of Perinatology* 30, (2010):155–162, http://www.nature.com/jp/journal/v30/n3/full/jp2009107a.html.

97   endometriosis: "Endometriosis," Mayo Clinic, http://www.mayoclinic.org/diseases-conditions/endometriosis/basics/definition/con-20013968, accessed December 5, 2015.

97   lower rates of polycystic ovary syndrome: Melanie Haiken, "How being a mom can make you healthier," Baby Center, http://www.babycenter.com/0_how-being-a-mom-can-make-you-healthier_1438536.bc?showAll=true, accesssed December 5, 2015.

97   uterine prolapse: Melanie Haiken, "Uterine Prolapse," Mayo Clinic, http://www.mayoclinic.org/diseases-conditions/uterine-prolapse/basics/definition/con-20027708, accessed December 5, 2015.

98   postpartum depression: "What is postpartum depression & anxiety?," American Psychological Association, http://www.apa.org/pi/women/resources/reports/postpartum-dep.aspx, accessed December 5, 2015.

98   first year after childbirth: Ariana Eunjung Cha, "It turns out parenthood is worse than divorce, unemployment—even the death of a partner," *Washington Post* (August 2015), https://www.washingtonpost.com/news/to-your-health/wp/2015/08/11/the-most-depressing-statistic-imaginable-about-being-a-new-parent/.

98   As you get older, your lungs: Gari Lesnoff-Caravaglia, *Health Aspects of Aging: The Experience of Growing Old* (Springfield, IL: Charles C. Thomas, 2007).

99   cancer is the second leading cause of death: Cancer Facts and Figures 2015, American Cancer Society, http://www.cancer.org/acs/groups/content/@editorial/documents/document/acspc-044552.pdf.

99   one-third of cancers: "Americans can prevent 1/3 of the most common cancers," American Institute for Cancer Research, http://www.aicr.org/learn-more-about-cancer/infographics-cancer-preventability.html, accessed December 7, 2015.

99   Lung cancer: "Tobacco-Related Cancers Fact Sheet," American Cancer Society, http://www.cancer.org/cancer/cancercauses/tobaccocancer/tobacco-related-cancer-fact-sheet.

99   Screening for cancer: "Screening Tests," National Cancer Institute, http://www.cancer.gov/about-cancer/screening/screening-tests, accessed November 28, 2015.

99   Heavy smokers, ages 55-80: "Should I get a lung CT scan?," Seattle Cancer Care Alliance, http://www.seattlecca.org/should-i-get-a-lung-ct-scan.cfm, accessed November 28, 2015.

99   get your flu shot: "Key Facts About Seasonal Flu Vaccine," CDC, http://www.cdc.gov/flu/protect/keyfacts.htm, accessed November 28, 2015.

100   focusing light through the lens: "How does the human eye work?," National Keratoconus Foundation, https://www.nkcf.org/how-the-human-eye-works/, accessed November 29, 2015.

100   Dry eye affects: A to Z Guide, NIH ORWH.

100   Spend time away from the bright light: "How to Maintain Good Eye Health," WebMD, http://www.webmd.com/eye-health/good-eyesight, accessed November 29, 2015.

100   diabetes and high blood pressure can also affect your eyes: Shereen Jegtvig, "How to Boost Your Diet and Nutrition to Protect Aging Eyes," allaboutvision.com, http://www.allaboutvision.com/over60/nutrition.htm, accessed November 29, 2015.

101   diagnosed with breast cancer: "Breast Cancer Facts," National Breast Cancer Foundation, Inc., http://www.nationalbreastcancer.org/breast-cancer-facts, accessed December 5, 2015.

101   decreasing since the 1990s: Ibid.

101   BRCA genes: "BRCA1 and BRCA2: Cancer Risk and Genetic Testing," National Cancer Institute, http://www.cancer.gov/about-cancer/causes-prevention/genetics/brca-fact-sheet, accessed December 5, 2015.

101   lifetime risk: Ibid.

101   all women should be offered testing: "A Never-Ending Genetic Quest," *New York Times* (February 2015), http://www.nytimes.com/2015/02/10/science/mary-claire-kings-pioneering-gene-work-from-breast-cancer-to-human-rights.html?_r=0, accessed December 1, 2015.

102   vitamins, over-the-counter drugs: "7 Secrets to Keeping Your Kidneys Healthy," Cleveland Clinic (April 2015), http://health.clevelandclinic.org/2015/04/7-secrets-to-keeping-your-kidneys-healthy/, accessed November 29, 2015.

102   hypertension increases with age: "Risk Factors for High Blood Pressure," National Heart Lung and Blood Institute (NHLBI).

102   risk factors for high blood pressure: Ibid.

103   Symptoms of heart attack for women: "Heart Attack Fact Sheet," NHLBI, http://www.nhlbi.nih.gov/files/docs/public/heart/heart_a2000 Nov ttack_fs_en.pdf.

103   47 percent of sudden cardiac deaths: "Heart Disease Facts," CDC, http://www.cdc.gov/heartdisease/facts.htm, accessed November 29, 2015.

104   increase your risk, including age, family history: "What Is Atherosclerosis?," National Heart Lung and Blood Institute (NHLBI), http://www.nhlbi.nih.gov/health/health-topics/topics/atherosclerosis, accessed November 29, 2015.

104   reduced the incidence of heart attacks: "Side Effects of Cholesterol-Lowering Statin Drugs," WebMD, http://www.webmd.com/cholesterol-management/side-effects-of-statin-drugs, accessed November 29, 2015.

104   potentially serious side effects: Ibid.

104   our entire skeleton is replaced: "Bone Health and Osteoporosis: A Report of the Surgeon General" (Rockville, MD: Office of the Surgeon General (US), 2004), http://www.ncbi.nlm.nih.gov/books/NBK45504/.

104   estrogen deficiency promotes the production of osteoclasts: B. Lawrence Riggs, "The mechanisms of estrogen regulation of bone resorption," Journal of Clinical Investigation 106, No. 10 (November 2000): 1203–1204, http://www.ncbi.nlm.nih.gov/pmc/articles/PMC381441/.

104   four times greater risk for developing osteoporosis than men: "Menopause & Osteoporosis," Cleveland Clinic, http://my.clevelandclinic.org/health/diseases_conditions/hic-what-is-perimenopause-menopause-postmenopause/hic_Menopause_and_Osteoporosis, accessed November 29, 2015.

104   Bone loss is increased by inactivity, low vitamin D: "Osteoporosis – overview," Medline Plus, U.S. NLM, http://www.nlm.nih.gov/medlineplus/ency/article/000360.htm, accessed November 29, 2015.

104 risk of broken bones: Ibid.

104 inhibiting osteoclasts: "Bisphosphonates," drugs.com, http://www.drugs.com/drug-class/bisphosphonates.html, accessed November 29, 2015.

106 Some elderly women even lose pubic hair: Roger Melick and H. Pincus Taft, "Observations on body hair in old people," *The Journal of Clinical Endocrinology and Metabolism* 19, No. 12 (December 1959): 1597-1607, http://press.endocrine.org/doi/abs/10.1210/jcem-19-12-1597.

106 up to four times longer to heal: "Aging changed in skin," Medline Plus, U.S. NLM, https://www.nlm.nih.gov/medlineplus/ency/article/004014.htm, accessed December 7, 2015.

106 more prone to sunburn: "Skin Problems & Treatments Health Center," WebMD, http://www.webmd.com/skin-problems-and-treatments/tc/sunburn-topic-overview, accessed December 7, 2015.

106 women develop hair: "Hirsutism," Mayo Clinic, http://www.mayoclinic.org/diseases-conditions/hirsutism/basics/definition/con-20028919 accessed November 29, 2015.

106 top-heavy testosterone-to-estrogen level: "Hirsutism—causes,"NHS Choices, http://www.nhs.uk/Conditions/hirsutism/Pages/causes.aspx, accessed November 29, 2015.

106 inheritance of diseases: "Hirsutism," Mayo Clinic.

106 medical treatments, including a cream: Ibid.

107 our stomach becomes less elastic: "Special Report: Gut feelings; A look at digestive health problems," Supplement to *Mayo Clinic Health Letter* (June 2012): www.healthletter.mayoclinic.com/DigestiveHealthSR.pdf.

107 liver loses cells with age: Ibid.

107 If you don't drink enough water: Ibid.

107 bone marrow, which makes blood cells: "Immune and lymphatic systems," InnerBody, http://www.innerbody.com/image/lympov.html#full-description, accessed November 29, 2015.

108 With age, our immune system begins to slow down: "Aging changes in immunity," Medline Plus, U.S. NLM, https://www.nlm.nih.gov/medlineplus/ency/article/004008.htm, accessed November 29, 2015.

108 The flu vaccine: Katherine Goodwin et al., "Antibody response to influenza vaccination in the elderly: a quantitative review," *Vaccine*, 24, No. 8 (February 2006):1159-69, http://www.ncbi.nlm.nih.gov/pubmed/16213065.

108 immune-system response to the flu: Joan B. Mannick, "mTOR inhibition improves immune function in the elderly," *Science Translational Medicine* 6, No. 268 (December 2014): 268ra179, http://stm.sciencemag.org/content/6/268/268ra179.

108 Get your flu shot every year: "What You Should Know and Do this Flu Season If You Are 65 Years and Older," CDC, http://www.cdc.gov/flu/about/disease/65over.htm, accessed December 7, 2015.

CHAPTER 8

112 as early as 42: "Evidence-based Practice Center Systematic Review Protocol: Menopausal Symptoms: Comparative Effectiveness Review of Therapies,"Agency for Healthcare Research and Quality (June 2013), http://effectivehealthcare.ahrq.gov/ehc/products/353/1022/menopause-protocol-130612.pdf.

114   later-onset menopause, which has been linked to longevity: Ellen B. Gold, "The Timing of the Age at Which Natural Menopause Occurs", Obstetrics and Gynecology Clinics of North America 38, No. 3 (September 2011): 425–440, http://www.ncbi.nlm.nih.gov/pmc/articles/PMC3285482/

114   increased risk of breast, endometrial, and ovarian cancer: Ibid.

114   Hormone replacement can be: "Definition of Surgical menopause," MedicineNet.com WebMD, http://www.medicinenet.com/script/main/art.asp?articlekey=8952, accessed November 29, 2015.

114   heart disease: Rogerio A. Lobo, "Surgical menopause and cardiovascular risks," Menopause 14, No.3 Pt 2 (May-June 2007): 562-6, http://www.ncbi.nlm.nih.gov/pubmed/17476145.

115   later-onset menopause: Ellen B. Gold, "The Timing of the Age at Which Natural Menopause Occurs," Obstetrics and Gynecology Clinics of North America 38, No. 3 (September 2011): 425–440, http://www.ncbi.nlm.nih.gov/pmc/articles/PMC3285482/.

115   increased risk of breast, endometrial: Ibid.

115   age at which your mother's: Stephanie Faris, "Late-Onset Menopause: What is Causing Your Delay?," Healthline Networks, Inc. (February 2012), http://www.healthline.com/health/menopause/late-onset#1, accessed November 29, 2015.

116   the Study of Women's Health across the Nation: "SWAN: Study of Women's Health across the Nation," SWAN study, http://www.swanstudy.org/, accessed December 8, 2015.

116   Stress, genetics: Beth A. Prairie, "Symptoms of depressed mood, disturbed sleep, and sexual problems in midlife women: cross-sectional data from the Study of Women's Health Across the Nation," Journal of Women's Health 24, No.2, (February 2015):119-26,http://www.ncbi.nlm.nih.gov/pubmed/25621768.

118   most anxious got the most severe: Nancy E. Avis, "Duration of Menopausal Vasomotor Symptoms Over the Menopause Transition," JAMA Internal Medicine 175, No. 4 (April 2015):531-9, http://www.ncbi.nlm.nih.gov/pubmed/25686030.

119   greatest risk for becoming depressed: Stacey B. Gramann, "Menopause and Mood Disorders," Medscape, WebMD, http://emedicine.medscape.com/article/295382-overview.

119   Depression raises your risk of heart disease: Abraham A. Ariyo, "Depressive Symptoms and Risks of Coronary Heart Disease and Mortality in Elderly Americans," Circulation 102 (October 2000): 1773-1779, http://circ.ahajournals.org/content/102/15/1773.full.pdf+html.

119   heart disease affects the onset of menopause: Helen S. Kok et al., "Heart Disease Risk Determines Menopausal Age Rather Than the Reverse," Journal of the American College of Cardiology 47, No. 10 (May 2006).

119   affects your risk of various other diseases: Ellen B. Gold, "The Timing of the Age at Which Natural Menopause Occurs," Obstetrics and Gynecology Clinics of North America 38, No. 3 (September 2011): 425–440, http://www.ncbi.nlm.nih.gov/pmc/articles/PMC3285482/.

120   one-third to two-thirds of women experience brain fog during menopause: "'Brain Fog' of Menopause Confirmed," University of Rochester Medical Center (March 2012) https://www.urmc.rochester.edu/news/story/3436/brain-fog-of-menopause-confirmed.aspx.

120   Dr. Miriam Weber: Ibid.

121   38 percent of menopausal women: Howard M. Kravitz et al., "Sleep difficulty in women at midlife: a community survey of sleep and the menopausal transition," Menopause 10, No. 1 (January-February 2003):19-28, http://www.ncbi.nlm.nih.gov/pubmed/12544673.

121 relationship between sleep and menopause: Howard M. Kravitz and Hadine Joffe, "Sleep
     During the Perimenopause: A SWAN Story," *Obstetrics and Gynecology Clinics of North
     America* 38, No. 3 (September 2011): 567–586.

123 smoking, lower education levels menopause onset: Ellen B. Gold et al., "Factors Associated
     with Age at Natural Menopause in a Multiethnic Sample of Midlife Women," *American
     Journal of Epidemiology* 153, No. 9 (May 2001): 865-874, http://aje.oxfordjournals.org/
     content/153/9/865.full.

123 vasomotor symptoms are reported more frequently: Ellen B. Gold et al., "Longitudinal
     Analysis of the Association Between Vasomotor Symptoms and Race/Ethnicity Across the
     Menopausal Transition: Study of Women's Health Across the Nation," *American Journal
     of Public Health* 96, No. 7 (July 2006): 1226–1235, http://www.ncbi.nlm.nih.gov/pmc/
     articles/PMC1483882.

123 some Hispanic women: Robin R. Green et al., "Menopausal symptoms within a His-
     panic cohort: SWAN, the Study of Women's Health Across the Nation," *Womens Health
     (Lond Engl)* 5, No. 2 (March 2009): 127–133, http://www.ncbi.nlm.nih.gov/pmc/articles/
     PMC3270699/.

123 except for Asian women: Eun-Ok Im, "Menopausal Symptoms Among Four Major Ethnic
     Groups in the U.S.," *Western Journal of Nursing Research* 32, No. 4 (June 2010): 540–565,
     http://www.ncbi.nlm.nih.gov/pmc/articles/PMC3033753/#R54.

123 A 2012 study of healthy women: Susan E. Trompeter, et al., "Sexual Activity and Satisfac-
     tion in Healthy Community-dwelling Older Women," *American Journal of Medicine* 125,
     No. 1 (January 2012): 37-43.e1, http://www.ncbi.nlm.nih.gov/pmc/articles/PMC3246190/.

124 after a woman's final menstrual period: "Hormones & Menopause: Tips from the National
     Institute on Aging," NIA (August 2012), https://www.nia.nih.gov/health/publication/
     hormones-and-menopause.

125 heart disease, breast cancer, and stroke: Ibid.

125 cellular inflammation: Flora Engelmann and Ilhem Messaoudi, "The Impact of Meno-
     pause on Immune Senescence," *Open Longevity Science* 6 (2012): 101-111,
     http://benthamopen.com/ABSTRACT/TOLSJ-6-101.

126 more friends women had, the healthier: Randy Kamen, "A Compelling Argument About
     Why Women Need Friendships," Huffington Post (November 2012),
     http://www.huffingtonpost.com/randy-kamen-gredinger-edd/female-friend-
     ship_b_2193062.html.

126 women with breast cancer who have close relationships: Tom Valeo, "Good Friends Are
     Good for You," WebMD, http://www.webmd.com/balance/features/good-friends-are-
     good-for-you, accessed November 29, 2015.

CHAPTER 9

134 Physical activity strengthens the parts of the brain: Heidi Godman, "Regular exercise
     changes the brain to improve memory, thinking skills," Harvard Health Publications
     (April 2014), http://www.health.harvard.cdu/blog/regular-exercise-changes-brain-
     improve-memory-thinking-skills-201404097110, accessed December 5, 2015.

134 BDNF: Kristin L. Szuhany et al., "A meta-analytic review of the effects of exercise on
     brain-derived neurotrophic factor," *Journal of Psychiatric Research* 60, (January 2015):
     56-64, https://www.ncbi.nlm.nih.gov/pubmed/25455510.

134 may protect against cognitive decline: "More Evidence that Omega-3 Fatty Acids Support Brain Health," UC Irvine Institute for Memory Impairments and Neurological Disorders, http://www.alz.uci.edu/more-evidence-that-omega-3-fatty-acids-support-brain-health/, accessed November 29, 2015.

134 washes itself of sticky plaque build-up: Yasmin Anwar, "Poor sleep linked to toxic buildup of Alzheimer's protein, memory loss," *Berkeley News,* University of California, Berkeley (June 2015), http://news.berkeley.edu/2015/06/01/alzheimers-protein-memory-loss/; Jeff Iliff, "One more reason to get a good night's sleep," TedMed Talks 2014, http://tedmed.com/talks/show?id=293015; Emily Underwood, "Sleep: The Ultimate Brainwasher?," *Science* magazine (October 2013), http://news.sciencemag.org/brain-behavior/2013/10/sleep-ultimate-brainwasher, accessed November 30, 2015.

134 Getting adequate sleep: Matt Wood, "Sleep More, Eat Less: Improving Sleep Habits Can Lead to Healthier Food Choices," *Science Life*, University of Chicago Medicine and Biological Sciences (June 2014), http://sciencelife.uchospitals.edu/2014/06/12/sleep-more-eat-less-improving-sleep-habits-can-lead-to-healthier-food-choices/.

134 soda has been linked to shortened telomeres: Jeffrey Norris, "Sugared Soda Consumption, Cell Aging Associated in New Study," (October 2014), UCSF, https://www.ucsf.edu/news/2014/10/119431/sugared-soda-consumption-cell-aging-associated-new-study, accessed November 30, 2015.

135 The queen is unique: Silvia C. Remolina and Kimberly A. Hughes, "Evolution and mechanisms of long life and high fertility in queen honey bees," *AGE* 30 (September 2008):177–185, http://www.ncbi.nlm.nih.gov/pmc/articles/PMC2527632/#!po=1.47059.

135 she who gets the most royal jelly gets the crown: "The Colony and Its Organization," Mid-Atlantic Apiculture Research and Extension Colony, https://agdev.anr.udel.edu/maarec/honey-bee-biology/the-colony-and-its-organization/, accessed November 30, 2015.

135 good nutrition can extend life: Nicole Jankovic et al., "Adherence to a Healthy Diet According to the World Health Organization Guidelines and All-Cause Mortality in Elderly Adults From Europe and the United States," *American Journal of Epidemiology* 180, No. 10 (October 2014): 978-988.

136 heart disease, colon cancer, and stroke: "Mediterranean diet may lower stroke risk, study finds," Medical Xpress (February 2015), http://medicalxpress.com/news/2015-02-mediterranean-diet.html#inlRlv, accessed November 30, 2015.

136 improved cognitive health: Walter C. Willett et al., "Mediterranean diet pyramid: a cultural model for healthy eating," *American Journal of Clinical Nutrition* 61, No. 6 (1995): 1402-6S.

136 reduced their risk of developing Alzheimer's: Martha Clare Morris et al., "MIND diet associated with reduced incidence of Alzheimer's disease," *Alzheimer's & Dementia* 11, No. 9, (September 2015): 1007–1014, http://www.alzheimersanddementia.com/article/S1552-5260%2815%2900017-5/abstract.

138 baby boomers … are overweight: Alexandra Sifferlin, "The Vast Majority of Baby Boomers Are Overweight or Obese," *Time* (July 2014), http://time.com/2945095/the-vast-majority-of-baby-boomers-are-overweight-or-obese/, accessed November 30, 2015.

138 more likely to become depressed: Wiley-Blackwell, "Obesity And Depression May Be Linked," ScienceDaily, (June 2008), http://www.sciencedaily.com/releases/2008/06/080602152913.htm.

138 promote malnutrition: Mary Hickson, "Malnutrition and ageing," *Postgraduate Medical Journal* 82 (January 2006):2–8, http://www.ncbi.nlm.nih.gov/pmc/articles/PMC2563720/.

139   general guidelines for calcium intake: "Calcium and Vitamin D: Important at Every Age," NIH, Osteoporosis and Related Bone Diseases National Resource Center (May 2015), http://www.niams.nih.gov/Health_Info/Bone/Bone_Health/Nutrition/, accessed November 30, 2015.

139   not to exceed more than 2,000 mg a day: "What's On Your Plate: Vitamins & Minerals," NIA, https://www.nia.nih.gov/health/publication/whats-your-plate/vitamins-minerals, , accessed November 30, 2015.

139   general guidelines for vitamin D: Ibid.

139   one billion people worldwide: James M. Greenblatt, "The Breakthrough Depression Solution: Psychological Consequences of Vitamin D Deficiency," *Psychology Today* (November 2011), https://www.psychologytoday.com/blog/the-breakthrough-depression-solution/201111/psychological-consequences-vitamin-d-deficiency, accessed November 30, 2015.

139   in mood disorders: Sue Penckofer et al., "Vitamin D and Depression: Where is all the Sunshine?," *Issues in Mental Health Nursing* 31, No. 6 (June 2010): 385-393, http://www.ncbi.nlm.nih.gov/pmc/articles/PMC2908269/.

140   American Heart Association recommends: "Know Your Fats," American Heart Association, http://www.heart.org/HEARTORG/Conditions/Cholesterol/PreventionTreatmentofHighCholesterol/Know-Your-Fats_UCM_305628_Article.jsp#.Vmd36spG9l0, accessed December 8, 2015.

140   3,200 mg of salt: Harvard Women's Health Watch, "Sodium, salt, and you," Harvard Medical School.

140   recommended daily allowance on food labels: "Sodium in Your Diet: Using the Nutrition Facts Label to Reduce Your Intake," U.S. Food and Drug Administration (FDA), http://www.fda.gov/Food/ResourcesForYou/Consumers/ucm315393.htm.

140   Women should aim to get 1.5 mg of B6: "Vitamins and Minerals," NIA, https://www.nia.nih.gov/health/publication/whats-your-plate/vitamins-minerals.

141   B12 deficiency: "Vitamin B12," Mayo Clinic, http://www.mayoclinic.org/drugs-supplements/vitamin-b12/evidence/hrb-20060243, accessed November 30, 2015.

141   Symptoms of B12 deficiency: Ibid.

141   Women over the age of fourteen: "Vitamin B12 dietary supplement fact sheet," NIH Office of Dietary Supplements, https://ods.od.nih.gov/factsheets/VitaminB12-HealthProfessional/, accessed November 30, 2015.141

141   Women over nineteen should get: "Folate dietary supplement fact sheet," NIH Office of Dietary Supplements, https://ods.od.nih.gov/factsheets/Folate-HealthProfessional/, accessed November 30, 2015.

141   60 percent of American women: "Overweight, obesity, and weight loss fact sheet," Office on Women's Health Organization, http://womenshealth.gov/publications/our-publications/fact-sheet/overweight-weight-loss.html#a, accessed November 30, 2015.

141   more likely to suffer from a disability: Eileen Rillamas-Sun et al., "Obesity and late-age survival without major disease or disability in older women," *JAMA Internal Medicine* 174, No. 1 (January 2014):98-106, http://www.ncbi.nlm.nih.gov/pubmed/24217806.

142   especially vulnerable to some of alcohol's harmful effects: A to Z Guide, ORWH, http://orwh.od.nih.gov/resources/sexgendcrhealth/index.asp, accessed November 30, 2015.

142   gastrointestinal enzyme: Whitley and Wesley, AFP, http://www.aafp.org/afp/2009/1201/p1254.html, accessed November 30, 2015.

142   into your bloodstream: Conversation with Dr. Janine Clayton, National Institutes of Health, December 2014.

142   women are more susceptible to alcoholic liver disease: "Alcohol Alert," NIH National Institute of Alcohol Abuse and Alcoholism (December 1999), http://pubs.niaaa.nih.gov/publications/aa46.htm, accessed November 30, 2015.

143   As we age, our bodies are less primed: Geoffrey Goldspink, "Age-Related Loss of Muscle Mass and Strength," *Journal of Aging Research* (March 2012): 158279, http://www.ncbi.nlm.nih.gov/pmc/articles/PMC3312297/.

143   fitness and aging expert: "Simon Melov, PhD," Buck Institute for Research on Aging, http://www.thebuck.org/melovLab, accessed November 30, 2015.

144   up to 1 in 3 hip fractures: Virtual Health Care Team, "Falls and Hip Fractures: Incidence of Falls and Associated Morbidity & Mortality," School of Health Professions, University of Missouri-Columbia, http://shp.missouri.edu/vhct/case4007/index.htm, accessed December 8, 2015.

144   three-quarters of the roughly 250,000 hip fractures: Hip Fractures Among Older Adults, CDC, http://www.cdc.gov/HomeandRecreationalSafety/Falls/adulthipfx.html, accessed November 30, 2015.

144   runners over fifty: Eliza F. Chakravarty et al., "Reduced disability and mortality among aging runners: a 21-year longitudinal study," *Archives Internal Medicine* 168, No. 15 (August 2008); 1638-46, http://www.ncbi.nlm.nih.gov/pubmed/18695077.

145   protective benefits for our mental ability: Ashley Carvalho et al., "Physical activity and cognitive function in individuals over 60 years of age: a systematic review," *Clinical Interventions in Aging* 9 (April 2014): 661–682, http://www.ncbi.nlm.nih.gov/pmc/articles/PMC3990369/.

145   70 percent of the energy: "Metabolism and weight loss: How you burn calories," Mayo Clinic, http://www.mayoclinic.org/healthy-lifestyle/weight-loss/in-depth/metabolism/art-20046508, accessed December 8, 2015.

145   RMR drops about 1 or 2 percent: Susan B. Roberts and Gerard E. Dallal, "Energy requirements and aging," Public Health Nutrition 8, No. 7A, (October 2005):1028–1036, http://www.ncbi.nlm.nih.gov/pubmed/16277818.145145

145   less lean body mass: S. P. Tzankoff and A. H. Norris, "Effect of muscle mass decrease on age-related BMR changes," *Journal of Applied Physiology* 43, No. 6, (December 1977): 1001-1006, http://jap.physiology.org/content/43/6/1001.145

145   does the decline of RMR: Marie-Pierre St-Onge and Dympna Gallagher, "Body composition changes with aging: The cause or the result of alterations in metabolic rate and macronutrient oxidation?," *Nutrition* 26, No. 2 (February 2010): 152–155, http://www.ncbi.nlm.nih.gov/pmc/articles/PMC2880224/.

146   VARICOSE VEINS: "Varicose veins," Mayo Clinic, http://www.mayoclinic.org/diseases-conditions/varicose-veins/basics/definition/con-20043474, accessed December 1, 2015.

146   risk factor for varicose veins: "Varicose veins: Risk factors," Mayo Clinic, http://www.mayoclinic.org/diseases-conditions/varicose-veins/basics/risk-factors/con-20043474.

146   best ways to prevent them: "Varicose veins: Lifestyle and home remedies," Mayo Clinic, http://www.mayoclinic.org/diseases-conditions/varicose-veins/basics/lifestyle-home-remedies/con-20043474.

147   get moving, and keep moving: Rachael van Pelt et al., "Age-related decline in RMR in physically active men: relation to exercise volume and energy intake," *American Journal of*

Physiology E281 (2001): 633-639; Rachael van Pelt et al., "Regular exercise and the age-related decline in resting metabolic rate in women," *Journal of Clinical Endocrinology and Metabolism* 82 (1997): 3208-3212.

148   adults with insomnia: "How does exercise help those with chronic insomnia?,"National Sleep Foundation, https://sleepfoundation.org/ask-the-expert/how-does-exercise-help-those-chronic-insomnia accessed December 1, 2015.

148   Mind-body techniques: Julia Schlam Edelman, "HMS Guide to Successful Sleep: Strategies for Women," Harvard Medical School (March 2013), http://www.harvardhealthbooks.org/wp-content/uploads/2013/03/SuccessfulSleepStrategiesForWomenSampleChapter.pdf.

148   stretching and moderate-intensity workouts: Shelley S. Tworoger et al., "Effects of a Yearlong Moderate-Intensity Exercise and a Stretching Intervention on Sleep Quality in Postmenopausal Women," *SLEEP* 26, No. 7, (2003): 830-836.

148   BDNF: Kristin L. Szuhany et al., "A meta-analytic review of the effects of exercise on brain-derived neurotrophic factor," *Journal of Psychiatric Research* 60, (January 2015): 56-64 https://www.ncbi.nlm.nih.gov/pubmed/25455510.

148   Up to 90 percent of us: Katherine Harmon, "Is Cellulite Forever?," *Scientific American* (May 2009), http://www.scientificamerican.com/article/is-cellulite-forever/.

148   fat cells accumulate and push up: "Cellulite," Mayo Clinic, http://www.mayoclinic.org/diseases-conditions/cellulite/basics/causes/con-20029901, accessed December 1, 2015.

148   decreased estrogen: Harmon, "Is Cellulite Forever?".

149   SWAN study reported substantial difficulties: MaryFran Sowers et al., "The association of menopause and physical functioning in women at midlife," *Journal of the American Geriatric Society* 49, No. 11 (November 2001): 1485-92, http://www.ncbi.nlm.nih.gov/pubmed/11890587.

150   Fifteen percent of women in the United States still smoke cigarettes: "Cigarette Smoking in the United States," CDC, http://www.cdc.gov/tobacco/campaign/tips/resources/data/cigarette-smoking-in-united-states.html, accessed December 1, 2015.

150   major cause of preventable death: "Smoking and tobacco use," CDC, http://www.cdc.gov/tobacco/data_statistics/fact_sheets/fast_facts/, accessed December 8, 2015.

150   Tobacco smoke is full of carcinogens: "Known and Probable Human Carcinogens," American Cancer Society, http://www.cancer.org/cancer/cancercauses/othercarcinogens/generalinformationaboutcarcinogens/known-and-probable-human-carcinogens accessed December 1, 2015.

151   EVEN TWENTY MINUTES: "Within 20 Minutes of Quitting," CDC, http://www.cdc.gov/tobacco/data_statistics/sgr/2004/posters/20mins/, accessed December 1, 2015; "Stop Smoking Recovery Timetable," whyquit.com, http://whyquit.com/whyquit/A_Benefits_Time_Table.html, accessed December 1, 2015.

152   clearing out the waste: Jeffrey Iliff, "One more reason to get a good night's sleep," TEDMED 2015, http://tedmed.com/talks/show?id=293015.

152   Aim for seven hours of sleep: "Study Ties 6-7 Hours of Sleep to Longer Life," *New York Times* (February 2002), http://www.nytimes.com/2002/02/15/us/study-ties-6-7-hours-of-sleep-to-longer-life.html; Daniel F. Kripke, "Mortality Associated With Sleep Duration and Insomnia,"*Archives of General Psychiatry* 59, No. 2 (February 2002):131-136.

153   twenty-five-hour cycle: Jurgen Aschoff, "Cireadian Rhythms in Man: A self-sustained oscillator with an inherent frequency underlies human 24-hour periodicity," UCSD (1965)

https://mechanism.ucsd.edu/teaching/F11/philbiology2011/aschoff.circadianrhythmsinman.
1965.pdf.

153 backs of their knees: Sandra Blakeslee, "Study Offers Surprise on Working of Body's Clock," *New York Times* (January 1998), http://www.nytimes.com/1998/01/16/us/study-offers-surprise-on-working-of-body-s-clock.html.

154 When that cycle is disrupted: Keith C. Summa and Fred W. Turek, "Chronobiology and Obesity: Interactions between Circadian Rhythms and Energy Regulation," *Advances in Nutrition* 5 (May 2014): 312S-319S, http://advances.nutrition.org/content/5/3/312S.full.

154 "social jet lag": Marc Wittmann et al., "Social jetlag: misalignment of biological and social time," *Chronobiology International: The Journal of Biological and Medical Rhythm Research* 23, No. 1-2 (2006):497-509, http://www.tandfonline.com/doi/abs/10.1080/07420520500545979?journalCode=icbi20#.VmeX9spG9l0.

154 Shift work: Josephine Arendt, "Shift work: coping with the biological clock," *Occupational Medicine* 60, No. 1 (2010): 10-20, http://occmed.oxfordjournals.org/content/60/1/10.full.

155 increased likelihood of developing cancer: "Painting, Firefighting and Shiftwork," IARC monographs on the evaluation of carcinogenic risks to humans, WHO International Agency for Research on Cancer 98 (2010), monographs.iarc.fr/ENG/Monographs/vol98/mono98.pdf.

155 reduce blue light exposure at night: Rosie Blau, "The light therapeutic," *The Economist Intelligent Life* (May/June 2014), http://moreintelligentlife.com/content/features/rosie-blau/light-and-health#_.

156 cortex shrinks with age: Mo Costandi, "Why Poor Sleep and Forgetfulness Plague the Aging Brain," *Scientific American* (January 2013), http://www.scientificamerican.com/article/why-poor-sleep-and-forgetfulness-plague-the-aging-brain/.

156 Harvard Nurses' study: Elizabeth E. Devore, "Sleep Duration in Midlife and Later Life in Relation to Cognition," *Journal of the American Geriatrics Society* 62, No. 6 (June 2014): 1073–1081, http://onlinelibrary.wiley.com/doi/10.1111/jgs.12790/abstract.

156 SLEEP FOR CALM: Rosalind D. Cartwright, *The Twenty-four Hour Mind: The Role of Sleep and Dreaming in Our Emotional Lives* (New York: Oxford University Press, Inc., 2010), http://www.amazon.com/The-Twenty-four-Hour-Mind-Emotional/dp/0199896283.

156 SLEEP FOR BEAUTY: "Beauty Sleep: 5 Benefits for Your Skin", WebMD, http://www.webmd.com/beauty/skin/beauty-sleep?page=2, accessed December 1, 2015.

CHAPTER 10

160 accumulation of belly fat: Colette Bouchez, "Can Stress Cause Weight Gain?," WebMD, http://www.webmd.com/diet/can-stress-cause-weight-gain, accessed December 9, 2015.

160 make your hair turn gray: Lizette Borreli, "What Causes Gray Hair? The Influence of Genetics and Other Factors on Hair Color," Medical Daily (March 2015), http://www.medicaldaily.com/pulse/what-causes-gray-hair-influence-genetics-and-other-factors-hair-color-324622, accessed December 9, 2015.

161 Stress hormones can damage melanin cells: Coco Ballantyne, "Fact or Fiction?: Stress Causes Gray Hair," *Scientific American* (October 2007) http://www.scientificamerican.com/article/fact-or-fiction-stress-causes-gray-hair/.

161   stress hormones: Bruce S. McEwan, " Central effects of stress hormones in health and disease: understanding the protective and damaging effects of stress and stress mediators," *European Journal of Pharmacology* 583, No. 2-3 (April 2008): 174–185.

161   Chronic stress accelerates inflammaging: Nicole D. Powell et al., "Social stress up-regulates inflammatory gene expression in the leukocyte transcriptome via ß-adrenergic induction of myelopoiesis," *PNAS* 110, no. 41 (October 2013):16574–16579; "Seniors," *The American Institute of Stress,* http://www.stress.org/seniors/, accessed December 9, 2015.

163   THE ANATOMY OF YOUR IMMUNITY: "Components of the Immune System," WebMD, http://www.webmd.com/a-to-z-guides/components-of-the-immune-system, accessed December 2, 2015.

163   three pounds of foreign bacteria: "NIH Human Microbiome Project defines normal bacterial makeup of the body," NIH (2012), http://www.nih.gov/news-events/news-releases/nih-human-microbiome-project-defines-normal-bacterial-makeup-body, accessed December 9, 2015.

163   10 percent human: Ibid.

163   breastfeeding: Deborah L. Wingard, "Is breast-feeding in infancy associated with adult longevity?," *American Journal of Public Health* 84, No. 9 (September 1994):1458–1462., http://www.ncbi.nlm.nih.gov/pmc/articles/PMC1615186/.

165   acute inflammation: "Acute Inflammation," University of Washington, http://courses.washington.edu/conj/inflammation/acuteinflam.htm, accessed December 4, 2015.

166   cellular aging, or inflammaging: Claudio Franceschi and Judith Campisis, "Chronic inflammation (inflammaging) and its potential contribution to age-associated diseases," *Journal of Gerontology* 69, Suppl. 1 (June 2014):S4-9, http://www.ncbi.nlm.nih.gov/pubmed/24833586; Daniel Baylis et al., "Understanding how we age: insights into inflammaging," *Longevity and Healthspan* 2 (May 2013); http://www.longevityandhealthspan.com/content/2/1/8.

166   Chronic, systemic inflammation: Brent Bauer, "Buzzed on inflammation," Mayo Clinic Health Letter, http://healthletter.mayoclinic.com/editorial/editorial.cfm/i/163/t/Buzzed%20on%20inflammation accessed December 9, 2015; "Systemic Inflammation: A Driver of Neurodegenerative Disease?," AlzForum, http://www.alzforum.org/news/conference-coverage/systemic-inflammation-driver-neurodegenerative-disease.

166   c-reactive protein (CRP) is a biomarker in the blood: "C-reactive protein test," Mayo Clinic, http://www.mayoclinic.org/tests-procedures/c-reactive-protein/basics/definition/prc-20014480, accessed December 4, 2015.

166   interleukin-6 has been linked to the age-related illnesses: Marcello Maggio et al., "Interleukin-6 in Aging and Chronic Disease: A Magnificent Pathway," *Journal of Gerontology* 61, No. 6 (June 2006): 575-584, http://www.ncbi.nlm.nih.gov/pmc/articles/PMC2645627/.

167   digest her milk: http://www.ncbi.nlm.nih.gov/pubmed/17224259.

167   one thousand different species: Mamm Genome (2014) 25:49–74 DOI 10.1007/s00335-013-9488-5; *The microbiome: stress, health and disease,* Rachel D. Moloney, Lieve Desbonnet, Gerard Clarke, Timothy G. Dinan, John F. Cryan.

167   contribute to malnutrition: Noah Voreades et al., "Diet and the development of the human intestinal microbiome," *Frontiers in Microbiology* 5 (September 2014):494, http://journal.frontiersin.org/article/10.3389/fmicb.2014.00494/full.

168   gut bacteria: http://genomemag.com/change-your-microbiome-change-yourself/#.VP0gMFPF9oE.

168   inflammaging is the result: Biagi et al, 2010; http://journals.plos.org/plosone/article?id=10.1371/journal.pone.0010667.

168   frailty: Van Tongeren et al., 2005, http://www.ncbi.nlm.nih.gov/pmc/articles/PMC1265947/; Claesson et al., 2011 http://www.pnas.org/content/108/Supplement_1/4586.long.

168   diverse and nutritious diet: Ibid.

168   Clostridium difficile: http://www.webmd.com/digestive-disorders/clostridium-difficile-colitis.

168   they got better: http://www.ncbi.nlm.nih.gov/pmc/articles/PMC3223289/.

169   eating processed foods: Noah Voreades et al., "Diet and the development of the human intestinal microbiome," *Frontiers in Microbiology* 5 (September 2014), http://journal.frontiersin.org/article/10.3389/fmicb.2014.00494/full.

169   smoking: "Cigarette smoke changes the gut microbiome," American Microbiome Institute (June 2015), http://www.microbiomeinstitute.org/blog/2015/6/11/cigarette-smoke-changes-the-gut-microbiome.

169   chronically stressed-out: "Stress affects the balance of bacteria in the gut and immune response," ScienceDaily (March 2011), http://www.sciencedaily.com/releases/2011/03/110321094231.htm.

169   diversity in our microbiome: From recommendations by Dr. Dale Bredesen, the Buck Institute.

169   exercising: Mandy Oaklander, "The Happy Effect Exercising Has on Your Gut Bacteria," *Prevention,* http://www.prevention.com/health/healthy-living/exercise-makes-your-gut-bacteria-more-diverse, accessed December 8, 2015.

169   Eating a diverse range: http://journal/frontiersin.org/article/10.3389/fmicb.2014.00494/full.

169   in the face of stress: http://www.nytimes.com/2015/06/28/magazine/can-the-bacteria-in-your-gut-explain-your-mood.html.

169   Prebiotics are food: http://www.molecularneurodegeneration.com/content/9/1/36.

169   Processed and packaged foods use antimicrobial preservatives: P. Michael Davidson, "Food additive: Food processing," Encyclopedia Britannica, http://www.britannica.com/topic/food-additive.

169   microbial populations: http://www.ncbi.nlm.nih.gov/pubmed/21040780.

170   eating fermented foods: "Brain maker foods," David Perlmutter, M.D., http://www.drperlmutter.com/eat/brain-maker-foods/.

171   to your brain and back: Mamm Genome (2014) 25:49–74 DOI 10.1007/s00335-013-9488-5 *The microbiome: stress, health and disease,* Rachel D. Moloney, Lieve Desbonnet, Gerard Clarke, Timothy G. Dinan, John F. Cryan.

171   gut to your brain: https://www.psychologytoday.com/blog/the-athletes-way/201405/how-does-the-vagus-nerve-convey-gut-instincts-the-brain.

171   fear and anxiety: http://www.jneurosci.org/content/34/21/7067.

171   behavior and mood: http://www.nytimes.com/2015/06/28/magazine/can-the-bacteria-in-your-gut-explain-your-mood.html.

171   stress-related diseases of the central nervous system: Ibid.

172   cells create inflammation: Franceschi and Campisi, *Journal of Gerontology* .

172   levels of stress hormones: http://www.takingcharge.csh.umn.edu/enhance-your-wellbeing/environment/nature-and-us/how-does-nature-impact-our-wellbeing.

172   study of 200 breast cancer survivors: "Yoga can lower fatigue, inflammation in breast cancer survivors," ScienceDaily (January 2014): http://www.sciencedaily.com/releases/2014/01/140127164408.htm.

172   practice yoga frequently: http://news.nationalgeographic.com/news/2014/02/140207-yoga-cancer-inflammation-stress/.

172   Massage: http://www.mayoclinic.org/healthy-lifestyle/stress-management/in-depth/massage/art-20045743.

172   Aromatherapy: http://www.ncbi.nlm.nih.gov/pubmed/21854199.

172   Laugh: http://www.mayoclinic.org/healthy-lifestyle/stress-management/in-depth/stress-relief/art-20044456.

173   lengthening your telomeres: http://www.ls.ucdavis.edu/dss/news-and-research/shamatha-project-nov10.html.

173   promote longevity: Ibid.

CHAPTER 11

178   like getting dressed: http://www.merckmanuals.com/home/brain_spinal_cord_and_nerve_disorders/brain_dysfunction/brain_dysfunction_by_location.html.

178   temporal lobes: Juebin Huang, "Overview of Cerebral Function," Merck Manuals, http://www.merckmanuals.com/professional/neurologic-disorders/function-and-dysfunction-of-the-cerebral-lobes/overview-of-cerebral-function.

178   occipital lobes: Ibid.

178   frontal lobes: Ibid.

178   special cells that triangulate space: "The 2014 Nobel Prize in Physiology or Medicine-Press Release," Nobel Media AB 2014, http://www.nobelprize.org/nobel_prizes/medicine/laureates/2014/press.html, accessed December 9, 2015.

178   neurons that hold your memories: Roxanne Khamsi, "Jennifer Aniston strikes a nerve," *Nature News* (June 2005), http://www.nature.com/news/2005/050620/full/news050620-7.html.

179   replicate one second: http://www.cnet.com/news/fujitsu-supercomputer-simulates-1-second-of-brain-activity/.

179   gray and lumpy:http://www.ninds.nih.gov/disorders/brain_basics/know_your_brain.htm.

179   neurons that make up your brain have three parts: Ibid.

180   Neuron: http://www.nlm.nih.gov/medlineplus/ency/imagepages/18117.htm: 000http://www.ninds.nih.gov/disorders/brain_basics/know_your_brain.htm.

180   only one-third the size: "Growth of newborn babies' brains tracked," NHS (August 2014) http://www.nhs.uk/news/2014/08August/Pages/growth-of-newborn-babies-brains-tracked.aspx, accessed December 9, 2015.

181   fully mature physically: http://www.npr.org/templates/story/story.php?storyId=141164708, accessed December 8, 2015.

181   scans of teenagers: http://brainconnection.brainhq.com/2013/03/20/decision-making-is-still-a-work-in-progress-for-teenagers/.000   http://www.ncbi.nlm.nih.gov/pmc/articles/PMC2892678/.

181   Phineas Gage: http://www.smithsonianmag.com/history/phineas-gage-neurosciences-most-famous-patient-11390067/?no-ist.

182   gaze more deeply: http://www.ninds.nih.gov/disorders/brain_basics/know_your_brain.htm.

183   day you were born: http://www.livescience.com/33179-does-human-body-replace-cells-seven-years.html.

183   the hippocampus: http://www.scholarpedia.org/article/Adult_neurogenesis.

183   amount of glial cells: http://blogs.scientificamerican.com/brainwaves/2012/05/16/know-your-neurons-classifying-the-many-types-of-cells-in-the-neuron-forest/.

184   Brain imaging studies: Rachel Marsh, "Neuroimaging Studies of Normal Brain Development and Their Relevance for Understanding Childhood Neuropsychiatric Disorders," *Journal of the American Academy of Child and Adolescent Psychiatry* 47, No. 11 (November 2008): 1233–1251, http://www.ncbi.nlm.nih.gov/pmc/articles/PMC2759682/.

184   Dr. Shennan Weiss: "Shennan Weiss, M.D., Ph.D.," Department of Neurology, UCLA, http://neurology.ucla.edu/directory/fellows/shennanweiss.

184   old dogs *can* learn new tricks: Carolyn Gregoire, "Old Brains Become Young Again in Neuroplasticity Study," The Huffington Post (May 2015), http://www.huffingtonpost.com/2015/05/22/brain-aging-neuroplasticity_n_7307662.

184   "cognitive reserve": http://www.ncbi.nlm.nih.gov/pubmed/23079557.

185   build stronger brains by challenging: Tim Adams, "Norman Doidge: the man teaching us to change our minds," *The Observer* (February 2015), http://www.theguardian.com/science/2015/feb/08/norman-doidge-brain-healing-neuroplasticity-interview.

185   Bassett's work: http://www.macfound.org/fellows/907/.

186   how Alzheimer's functions: Robert M. Koffie et al., "Alzheimer's disease: synapses gone cold," *Molecular Neurodegeneration* 6, No. 63 (August 2011): http://www.molecularneurodegeneration.com/content/6/1/63.

187   NEUROTRANSMITTERS: Much of the information on neurotransmitters in this section comes from http://www.ninds.nih.gov/disorders/brain_basics/know_your_brain.htm.

187   Many studies in rodents: http://www.sciencedirect.com/science/article/pii/S0047637400002256.

187   in the motor cortex: http://www.ncbi.nlm.nih.gov/pmc/articles/PMC2443746/.

187   muscle strength and movement: http://www.ncbi.nlm.nih.gov/pubmed/21529329.

187   secrete hormones: http://www.ncbi.nlm.nih.gov/books/NBK11143/.

187   with cognitive decline: http://www.ncbi.nlm.nih.gov/pubmed/15312959.

187   acetylcholine levels: http://www.ncbi.nlm.nih.gov/pubmed/?term=nordberg+2001+alzheimer; http://www.ncbi.nlm.nih.gov/pubmed/15190684.

188   Low levels of GABA: http://www.ncbi.nlm.nih.gov/pubmed/21889518.

188   raise our GABA levels: http://medicine.jrank.org/pages/1225/Neurotransmitters-GABA-glutamate.html: http://well.blogs.nytimes.com/2013/07/03/how-exercise-can-calm-anxiety/?_r=0.

188   lessened anxiety: http://www.ncbi.nlm.nih.gov/pmc/articles/PMC3111147/.

188   "read" serotonin's message: http://www.sciencemag.org/content/226/4681/1393.

188   Dopamine is an excitatory neurotransmitter: John D. Salamone and Mercè Correa, "The Mysterious Motivational Functions of Mesolimbic Dopamine," *Neuron* 76, No. 3 (November 2012): 470–485, http://www.cell.com/neuron/abstract/S0896-6273 (12)00941-5.

188   Dopamine levels decrease as we age: D.F. Wong et al., "Effects of age on dopamine and serotonin receptors measured by positron tomography in the living human brain," *Science* 226, No. 4681 (December 1984): 1393-1396, http://www.ncbi.nlm.nih.gov/pubmedhealth/PMH0001762/.

188   dopamine loss in age: http://www.ncbi.nlm.nih.gov/pubmedhealth/PMH0001762/.

188   meditation can help to raise: http://www.ncbi.nlm.nih.gov/pmc/articles/PMC3044190/.

188   sixth-leading cause of death: http://www.alz.org/facts/.

188   two-thirds are women: Ibid.

188   forgetting can be a natural part of getting older: "Forgetfulness: Knowing When to Ask for Help," NIA, https://www.nia.nih.gov/health/publication/forgetfulness#age, accessed December 9, 2015.

189   eating habits and hygiene: http://memory.ucsf.edu/brain/aging/dementia.

189   $200 billion: http://www.alz.org/news_and_events_law_by_Obama.asp.

190   the MEND protocol: Dale E. Bredesen, "Reversal of cognitive decline: a novel therapeutic program," *Aging* 6, No. 9 (September 2014): 707-17, http://www.ncbi.nlm.nih.gov/pubmed/25324467.

190   plaques outside the cells: http://www.sciencedirect.com/science/article/pii/S0925443909002427.

191   sticky misfolded proteins: http://www.npr.org/sections/health-shots/2013/10/18/236211811/brains-sweep-themselves-clean-of-toxins-during-sleep.

191   social isolation: http://memory.ucsf.edu/brain/aging/overview; http://www.ncbi.nlm.nih.gov/pubmed/25956016.

191   decreased risk for cognitive decline: "Preventing Alzheimer's Disease: What Do We Know?," NIA.

192   increased volume of the hippocampus: Gretchen Reynolds, "Can Exercise Reduce Alzheimer's Risk?," *New York Times* (July 2014), http://well.blogs.nytimes.com/2014/07/02/can-exercise-reduce-alzheimers-risk/?_r=0 ,accessed December 9, 2015.

192   clearing out the harmful plaques: Hamilton, NPR.

192   new synaptic connections: Bill Hathaway, "Yale team discovers how stress and depression can shrink the brain," *Yale News* (August 2012) http://news.yale.edu/2012/08/12/yale-team-discovers-how-stress-and-depression-can-shrink-brain.

192   what it has lost: http://www.cell.com/neuron/abstract/S0896-6273%2813%2900544-8?_returnURL=http%3A%2F%2Flinkinghub.elsevier.com%2Fretrieve%2Fpii%2FS08966 27313005448%3Fshowall%3Dtrue.

193   lower levels of the stress hormone: "Mindfulness Meditation Could Lower Levels of Cortisol, the Stress Hormone," Huffington Post (March 2013), http://www.huffingtonpost.com/2013/03/31/mindfulness-meditation-cortisol-stress-levels_n_2965197.html, accessed December 10, 2015.

193   exhibited physical changes in gray matter: Sue McGreevey, "Eight weeks to a better brain," *Harvard Gazette* (January 2011), http://news.harvard.edu/gazette/story/2011/01/eight-weeks-to-a-better-brain/ accessed December 10, 2015.

193   reduced amygdala activity: Sue McGreevey, "Meditation's positive residual effects," *Harvard Gazette* (November 2012), http://news.harvard.edu/gazette/story/2012/11/meditations-positive-residual-effects/, accessed December 10, 2015.

193   monolingual adults: http://www.alzheimers.net/2013-11-11/speaking-two-languages-delays-dementia/; http://www.ncbi.nlm.nih.gov/pubmed/7730528.194

194   our brains are wired: http://www.scientificamerican.com/article/music-changes-the-way-you-think/.

194   when people listen to music: "Listening to music lights up the whole brain," Science Daily (December 2011), http://www.sciencedaily.com/releases/2011/12/111205081731.htm; Mallika Rao, "Playing Music Gives Brains A 'Full Body Workout'," Says Science," Huffington Post (November 2014), http://www.huffingtonpost.com/2014/11/07/playing-music-brain-workout_n_6116546.html.

194   People who play music regularly: "Early music lessons boost brain development," Science Daily (February 2013).

194   Music can also help with memory: "5 Reasons Why Music Boosts Brain Activity," Alzheimers.net (July 2014) http://www.alzheimers.net/2014-07-21/why-music-boosts-brain-activity-in-dementia-patients/.

194   changes in cognition will be a part of healthy aging: "Forgetfulness: Knowing When to Ask for Help," NIA.

CHAPTER 12

197   increased risk of mortality:http://www.pnas.org/content/110/15/5797.full; http://assets.aarp.org/rgcenter/general/loneliness_2010.pdf.

198   to die prematurely: http://journals.plos.org/plosmedicine/article?id=10.1371/journal.pmed.1000316.

198   people who volunteer: http://hpq.sagepub.com/content/10/6/739.short.

199   lower levels of inflammatory markers: Elliot Friedman et al., "Plasma inerleukin-6 and soluble IL-6 receptors are associated with psychological well-being in aging women," *Health Psychology* 26, No. 3 (May 2007): 305-313.

199   People with regular social ties: http://www.alzprevention.org/lifestyle-choices-about-socialization.php.

200   longer telomere length: Eli Puterman and Elissa Epel, "An intricate dance: Life experience, multisystem resiliency, and rate of telomere decline throughout the lifespan," *Social and Personality Psychology Compass* 6, No. 11 (November 2012): 807–825, http://www.ncbi.nlm.nih.gov/pmc/articles/PMC3496269/.

200   stronger immne system: http://www.mayoclinic.org/healthy-lifestyle/stress-management/in-depth/stress-relief/art-20044456.

200   risk factors for cardiovascular: http://www.researchgate.net/profile/Wendy_Troxel2/publication/231585784_Marital_status_and_quality_in_middle-aged_women_Associations_with_levels_and_trajectories_of_cardiovascular_risk_factors/links/0046353b20fbaeba47000000.pdf.

200 study that observed 6,500 men and women: Andrew Steptoe et al., "Social isolation, lone-liness, and all-cause mortality in older men and women," PNAS 110, No. 15 (April 2013): 5797–5801, http://www.pnas.org/content/110/15/5797.abstract.

201 WOMEN AND DEPRESSION: https://www.nimh.nih.gov/health/topics/depression/index.shtml.

201 "visiting demon": http://www.latimes.com/entertainment/la-ca-kathleen-norris21-2008sep21-story.html, accessed December 1, 2015.

201 symptoms of clinical depression: http://www.mayoclinic.org/diseases-conditions/depression/expert-answers/clinical-depression/faq-20057770.

201 rise in depression with age: http://www.cdc.gov/aging/mentalhealth/depression.htm.

INDEX

Page numbers of illustrations appear in italics.

heart disease (*cont.*)

social isolation and, 200, 202

  statin drugs and, 104

  stem cell treatment and, 39

  surgery for heart disease, 31

  women and mortality from, 47

  women's heart attack symptoms, 47, 103

heart health, 29, 30–32, 38

  diet and nutrition, 140

  exercise and, 147

  seventies decade and, 69

  women's hearts, 46–47

high blood pressure (hypertension), 69

  ADH and, 92

  African Americans and, 102

  chronic inflammation, 166

  diastolic and systolic numbers, 102

  environmental factors, 102

  eye health and, 100

  heart disease and, 30

  kidney health and, 100

  prescription drugs and, 102

  risk factors, 102, 141

  the silent killer, 102

  social isolation and, 200, 202

  stress reduction and, 172

Hodes, Richard J., 131, 202

homeostasis, 91

hormones, 91–93, 96, 121, 124–25. *See also specific hormones*

  aging and, 90, 95–96

  cellular aging and, 87

  symptoms of declining levels of sex hormones, 96

hormone therapy, 114, 125

human genome, 82

Human Genome Project, 82

human growth hormone (HGH), 91–92

hypertension. *See* high blood pressure

hyperthyroidism, 93, 140

hypothyroidism, 92

hysterical neurosis, 50

I

immune system

  aging and, 163–65

  anatomy of, 163

  cancer and, 164–65

  development of, 162

  fifties decade and, 164

  how it ages, 107–8

  how to protect, 108

  laughing and, 172

  meditation and, 173

  microbiome and, 163, 167–68

  mother's milk and, 163–64

  nutrition and, 108

  senescent cells and, 86

  signs of weakened, 164

  sleep and, 108, 134

  social isolation and weakened, 202

  stress and, 160, 161–62, 165, 176

inflammation, 165–66

  aging acceleration and, 86–87

  biomarkers for, 166

  chronic and cellular aging (inflammaging), 166, 168

  free radicals and chronic, 78

  menopause and, 125

  runners and, 144

  social connection and, 199, 200

  stress and, 169, 172

  sugar consumption and, 134

inflammatory bowel disease, 166

influenza, 108, 164

injuries, 45

interleukin-6 (IL6), 166

International Agency for Research on Cancer (IARC), 155

Iris Cantor-UCLA Women's Research Center, 116

J

joint health, 69, 93, 166

K

Kennedy, Brian, 37

kidney health, 100, 102

King, Mary-Claire, 101

Knight and Day (film), 177

L

lactose intolerance, 107

learning, 6

  brain health and, 109, 193

  neural connections and, 185, 193

Levy, Becca, 72

Lieberman, Matthew D., 180–81

lifestyle choices. *See also* diet and nutrition; exercise; sleep; smoking; stress

  aging and, 23, 63, 64–65

  Cameron's grandparents, 61, 64, 65–66

  cellular aging and, 75, 85

  education and, 32

  genes, heredity, and, 65–66

ABOUT THE AUTHORS

CAMERON DIAZ has been telling stories as a film actor for more than two decades. She is also the author of the #1 *New York Times* bestseller, *The Body Book*, and an excellent cook. She supports numerous causes that advocate environmental concerns, education, and the empowerment of women and girls. Cameron lives with her husband and assorted animals in Los Angeles.

SANDRA BARK is a *New York Times* bestselling author who collaborates with smart, passionate people to turn ideas into books. Learn more about her work at www.sandrabark.com.